Executing Temporal Logic Programs

Executing Temporal Logic Programs

B. C. MOSZKOWSKI

Computer Laboratory, University of Cambridge

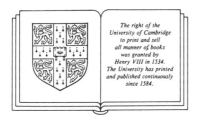

CAMBRIDGE UNIVERSITY PRESS
Cambridge
New York   New Rochelle
Melbourne   Sydney

Published by the Press Syndicate of the University of Cambridge
The Pitt Building, Trumpington Street, Cambridge CB2 1RP
32 East 57th Street, New York, NY 10022, USA
10 Stamford Road, Oakleigh, Melbourne 3166, Australia

© Cambridge University Press 1986

First published 1986
Reprinted 1987

Printed in Great Britain at the
University Press, Cambridge

British Library cataloguing in publication data

Moszkowski, B. C.
Executing temporal logic programs.

1. Electronic digital computers
-programming
I. Title
005.1   QA76.6

ISBN 0 521 31099 7

To Doody, Debbie and all the others

Contents

	Acknowledgements	xiii
1	**Introduction**	**1**
1.1	Organization of Presentation	2
2	**Basic Features of Temporal Logic**	**3**
2.1	Background	3
2.2	Syntax of the Logic	5
	2.2.1 Syntax of expressions	5
	2.2.2 Syntax of formulas	5
2.3	Models	6
2.4	Interpretation of Expressions and Formulas	7
2.5	Satisfiability and Validity	9
3	**Deriving Other Operators**	**10**
3.1	Boolean Operators	10
3.2	The Operator $\Diamond$	11
3.3	The Operators *empty* and *more*	12
3.4	The Operators *gets* and *stable*	12
3.5	The Operator *halt*	13
3.6	Temporal Equality	14
3.7	The Operator *weak next*	14
4	**Programming in Temporal Logic**	**16**
4.1	Syntax of Tempura	18
	4.1.1 Locations	18
	4.1.2 Expressions	18
	4.1.3 Statements	19

4.2	Some Other Statements	20
4.3	Determinism	20
5	**Some Additional Constructs**	**21**
5.1	Static Variables	21
5.2	Quantified Formulas	22
5.3	Enlarging the Data Domain	24
5.4	The Operator *len*	25
5.5	Bounded Quantification	26
5.6	The Operator *fin*	27
5.7	Temporal Assignment	27
5.8	Incorporating these Constructs into Tempura	29
	5.8.1 Locations	29
	5.8.2 Expressions	29
	5.8.3 Statements	30
6	**The Operator *chop***	**32**
6.1	Syntax and Semantics of *chop*	32
6.2	Discussion of the Operator *chop*	33
6.3	Simple For-Loops	34
6.4	Indexed For-Loops	35
6.5	While-Loops and Related Constructs	35
6.6	Deriving For-Loops and While-Loops	36
6.7	The Construct *skip*	38
6.8	Incorporating these Constructs into Tempura	39
7	**Some Applications**	**41**
7.1	Tree Summation	41
	7.1.1 Serial solution	42
	7.1.2 Parallel solution	43
	7.1.3 Correctness and performance	47
7.2	Partitioning a List	48
	7.2.1 Correctness of *partition_list*	48
7.3	Quicksort	51
	7.3.1 Explanation of *quick_partition*	53
	7.3.2 Explanation of *serial_sort_parts*	55
	7.3.3 Parallel quicksort	55

7.4	A Multiplication Circuit	55
7.5	Pulse Generation	57
7.6	Testing a Latch	61
7.7	Synchronized Communication	61

8 An Interpreter for Tempura — 69

8.1	Variables Used by the Interpreter	70
8.2	Basic Execution Algorithm	72
8.3	Description of the Procedure *execute_single_state*	73
8.4	Description of the Procedure *transform_stmt*	73
	8.4.1 Implementing the statements *true* and *false*	75
	8.4.2 Implementing equalities	75
	8.4.3 Implementing *empty* and *more*	76
	8.4.4 Implementing *request* and *display*	76
	8.4.5 Implementing conjunctions	77
	8.4.6 Transforming $\circledcirc w$, $\bigcirc w$ and $\square w$	77
	8.4.7 Implementing implication	77
	8.4.8 Implementing some other statements	77
8.5	Implementing Locations	77
8.6	Implementing Expressions	78
8.7	Static Variables, Lists and Quantifiers	79
	8.7.1 Implementing static variables	79
	8.7.2 Implementing lists	79
	8.7.3 Implementing *list*, *fixed_list* and *stable_struct*	79
	8.7.4 Implementing list constructors	80
	8.7.5 Implementing subscripts	80
	8.7.6 Implementing the list-length operator	81
	8.7.7 Implementing existential quantification	81
	8.7.8 Implementing universal quantification	82
	8.7.9 Implementing the operators *len*, *fin* and $\leftarrow$	82
	8.7.10 Implementing predicate and function definitions	82
	8.7.11 Implementing predicate and function invocations	83
8.8	Implementing *chop* and Iterative Operators	84
	8.8.1 Implementing the operator *chop*	84
	8.8.2 Implementing iterative operators	85

	8.8.3	Implementing the operator *skip*	86
8.9	Alternative Interpreters		86
	8.9.1	Immediate assignments	86
	8.9.2	Two-level memory	87
	8.9.3	Single-pass processing	87
	8.9.4	Time stamps	87
	8.9.5	Ignoring the operator *stable*	88
	8.9.6	Redundant assignments	88
	8.9.7	Special-purpose constructs	89
	8.9.8	Suppressing checks	89
	8.9.9	Call-by-name	89
9	**Experimental Features**		**91**
9.1	Temporal Projection		91
	9.1.1	Incorporating *proj* in Tempura	93
	9.1.2	Universal projection	94
9.2	Lambda Expressions and Pointers		96
	9.2.1	Lambda expressions	96
	9.2.2	Pointers	96
9.3	The *process* Construct		97
9.4	The *prefix* Construct		98
	9.4.1	Incorporating *prefix* in Tempura	98
9.5	Implementing these Constructs		102
	9.5.1	Implementing temporal projection	102
	9.5.2	Implementing lambda expressions and pointers	103
	9.5.3	Implementing the *process* construct	103
	9.5.4	Implementing the *prefix* construct	104
10	**Discussion**		**105**
10.1	Experience and Further Work		105
10.2	Related Programming Formalisms		106
	10.2.1	The programming language *Lucid*	106
	10.2.2	*CCS* and *CSP*	106
	10.2.3	Predicative programming	107
	10.2.4	The programming language *Esterel*	108
	10.2.5	The programming language *Prolog*	109

	10.2.6 Functional programming	111
10.3	Other Work on Temporal Logic	112
	10.3.1 Interval logic	113
	10.3.2 Generalized *next* operator	113
	10.3.3 Temporal logic as an intermediate language	114
	10.3.4 Semantics based on transition graphs	114
	10.3.5 Compositional proof rules	115
	10.3.6 Synthesis from temporal logic	116
	10.3.7 Automatic verification of circuits	118
10.4	Conclusions	118
	Bibliography	**120**

Acknowledgements

I would first and foremost like to express my deep gratitude to Mike Gordon. Not only did he initially propose that I come to Cambridge, but he also helped provide a stimulating and remarkably hassle-free environment in which this research could be pursued.

Roger Hale constructed a parser for Tempura and patiently served as the resident guinea pig and captive user during the development of the system. In addition, he critically read various drafts of this report.

Zohar Manna introduced me to temporal logic and its use in computer science. Tony Hoare was a ready source of encouragement and advice as this work developed. I also wish to thank Bill Clocksin, Miriam Leeser, Roger Needham, Keith van Rijsbergen and Edmund Ronald for their stimulating conversations and suggestions. A special mention should be made of Ernest Kirkwood at Cambridge University Press for his prompt response to the original manuscript of this book. Funding from the British Science and Engineering Research Council is gratefully acknowledged.

1 Introduction

Temporal logic [28,40] has been recently put forward as a useful tool for reasoning about concurrent programs and hardware. Within temporal logic, one can express logical operators corresponding to time-dependent concepts such as *"always"* and *"sometimes."* Consider, for example, the English sentence

"If the propositions P and Q are always true, then P is always true."

This can be represented in temporal logic by the formula

$$\Box(P \wedge Q) \supset \Box P.$$

Here the operator $\Box$ corresponds to the notion *"always."* Thus, the subformula $\Box(P \wedge Q)$ can be understood as *"P and Q are always true."*

Typically, temporal logic has been thought of as a tool for specifying and proving properties of programs written in, say, Hoare's CSP [22] or variants of Pascal with concurrency [20]. This distinction between temporal logic and programming languages has troubled us since it has meant that we must simultaneously use two separate notations. Programming formalisms such as Hoare logic [19], dynamic logic [15,39], and process logic [6,16] also reflect this dichotomy. For example, the following Hoare clause specifies that if I is initially 3, then after it increases by 1, its value is 4:

$$\{I = 3\}\ I := I + 1\ \{I = 4\}.$$

Here we have the formulas $I = 3$ and $I = 4$ as well as the statement $I := I + 1$.

One way to bridge the gap between logic and programs is by finding ways of using temporal logic itself as a tool for programming and simulation. With this in mind, we have developed *Tempura*, an imperative programming language based on subsets of temporal logic. Every Tempura statement is a temporal logic formula. This lets us specify and reason about Tempura programs without the need for two notations.

The underlying formalism used is called *Interval Temporal Logic* (ITL) [34,14,36] and includes such conventional temporal operators as $\bigcirc$ (*next*) and $\square$ (*always*) as well as lesser known ones such as *chop*. This provides a basis for the Tempura programming language. We present ITL and Tempura and describe several sample Tempura programs illustrating how to model the structure and behavior of hardware and software systems in a unified way. The design of an interpreter for Tempura is also discussed.

1.1 Organization of Presentation

We start off in chapter 2 by describing the syntax and semantics of a temporal logic having the operators $\square$ (*always*) and $\bigcirc$ (*next*). In chapter 3 a number of additional temporal constructs are derived and then used in chapter 4 to build legal Tempura programs. We extend ITL and Tempura in chapter 5 to include constructs for list structures as well as existential and universal quantifiers. The operator *chop* is introduced in chapter 6 and used to express for-loops and other iterative constructs. A variety of Tempura programming examples are then given in chapter 7 to show the utility of the language in dealing with hardware and software. After this is a discussion in chapter 8 on the details of implementing an interpreter for Tempura. In chapter 9 we investigate the concept of *temporal projection* as well as lambda expressions and pointers. Chapter 10 looks at the current status of work on Tempura interpreters and discusses future plans and related research.

2 Basic Features of Temporal Logic

Before describing Tempura, it is necessary to have an understanding of the underlying temporal logic. Some of the constructs described here are later used in Tempura programs. Others facilitate reasoning about program behavior. Rather than presenting the entire logic at once, we first introduce some basic operators. In later chapters, additional operators are considered.

2.1 Background

Let us first motivate the usage of temporal logic for specifying and reasoning about dynamic behavior. Readers who are already familiar with temporal logic can omit this discussion.

Predicate calculus [11] is a versatile and precise notation for formally specifying situations. For example, we can readily express the statement "*I equals 2 and J equals I plus 1*" by means of the following formula:

$$(I = 2) \wedge (J = I + 1).$$

However, dynamic behavior is more problematic. For instance, the statement "*The variable I at one time equals 1 and later equals 2*" is not satisfactorily handled by a formula such as

$$(I = 1) \vee (I = 2).$$

This just describes a static situation in which I could equal either 1 or 2. The formula

$$(I = 1) \wedge (I = 2)$$

is certainly not appropriate here since it is logically equivalent to *false*!

One way to get around the static nature of logic is by modelling time-dependent variables as explicit functions of time. For example, we might specify the changing values of I using the following formula:

$$\exists t, t': \bigl([t \leq t'] \wedge [I(t) = 1] \wedge [I(t') = 2] \bigr).$$

Here the variables t and t' indicate the two time points where I's values are examined. This technique for representing dynamic behavior is very powerful but suffers from the proliferation of extra time variables and quantifiers.

In this and subsequent sections we look at an alternative approach to reasoning about periods of time. We base it on temporal logic, a formalism that includes conventional logical operators such as $\wedge$ and $=$ as well as time-dependent ones such as $\square$ (read *"always"*) and $\Diamond$ (read *"sometimes"*). Although originally developed for application in philosophy, temporal logic has been put forward by Burstall [5], Pnueli [38] and others as a useful tool for dealing with computer programs and digital hardware.

Within the framework of temporal logic, it is possible to describe dynamic behavior in a simple and elegant fashion. For example, the statement *"I is always greater than 3 and sometimes less than 6"* can be expressed by means of the formula

$$\square(I > 3) \wedge \Diamond(I < 6).$$

The formula

$$\Diamond[(I = 1) \wedge \Diamond(I = 2)]$$

describes an interval of time in which the variable I at some time equals 1 and at some later time equals 2. Properties of time can also be expressed. For instance, if I always equals 1 and J sometimes equals 3 then we can infer that the sum $I+J$ sometimes equals 4:

$$[\square(I = 1) \wedge \Diamond(J = 3)] \supset \Diamond(I + J = 4).$$

These examples convey only a vague idea of the utility and convenience of temporal logic. As will be shown, temporal logic provides a natural means for describing such dynamic notations as stability, termination and interval length. Let us now look at the basic syntax and semantics of the formalism.

2.2 Syntax of the Logic

The initial set of constructs includes conventional logical operators such as $=$ (*equality*) and $\wedge$ (*logical-and*). In addition, there are the two temporal operators $\bigcirc$ (*next*) and $\square$ (*always*).

2.2.1 Syntax of expressions

Expressions are built inductively as follows:

- Individual variables: $A, B, C, \ldots$

- Functions: $f(e_1, \ldots, e_k)$, where $k \geq 0$ and $e_1, \ldots, e_k$ are expressions. In practice, we use functions such as $+$ and mod. Constants such as 0 and 1 are treated as zero-place functions.

- Next: $\bigcirc e$, where e is an expression.

Here are two examples of syntactically legal expressions:

$$I + (\bigcirc J) + 1, \qquad (\bigcirc I) + J - \bigcirc\bigcirc(I + \bigcirc J).$$

2.2.2 Syntax of formulas

Formulas are built inductively as follows:

- Predicates: $p(e_1, \ldots, e_k)$, where $k \geq 0$ and $e_1, \ldots, e_k$ are expressions. Predicates include $\leq$ and other basic relations.

- Equality: $e_1 = e_2$, where e_1 and e_2 are expressions.

- Logical connectives: $\neg w$ and $w_1 \wedge w_2$, where w, w_1 and w_2 are formulas.

- Next: $\bigcirc w$, where w is a formula.

- Always: $\square w$, where w is a formula.

Here are some syntactically legal ITL formulas:

$(J = 2) \wedge \bigcirc (I = 3)$,
$(\bigcirc \square [I = 3]) \wedge \neg ([\bigcirc J] = 4)$,
$\bigcirc (\square [I = 3] \wedge \bigcirc \bigcirc [J = 4])$.

Note that the operator $\bigcirc$ can be used both for expressions (e.g., $\bigcirc J$) and for formulas (e.g., $\bigcirc (I = 3)$).

2.3 Models

A model is a triple $(\mathcal{D}, \Sigma, \mathcal{M})$ containing a data domain $\mathcal{D}$, a set of states Σ and an interpretation $\mathcal{M}$ giving meaning to every function and predicate symbol. For the time being, we take the data domain $\mathcal{D}$ to be the integers. A state is a function mapping variables to values in $\mathcal{D}$. We let Σ be the set of all such functions. For a state s in Σ and a variable A, we let $s[\![A]\!]$ denote A's value in s. Each k-place function symbol f has an interpretation $\mathcal{M}[\![f]\!]$ which is a function mapping k elements in $\mathcal{D}$ to a single value:

$$\mathcal{M}[\![f]\!] \in (\mathcal{D}^k \to \mathcal{D}).$$

Interpretations of predicate symbols are similar but map to truth values:

$$\mathcal{M}[\![p]\!] \in (\mathcal{D}^k \to \{true, false\}).$$

We assume that $\mathcal{M}$ gives standard interpretations to operators such as $+$ and $<$.

The semantics given here keep the interpretations of function and predicate symbols independent of intervals. They can however be generalized to allow for functions and predicates that take into account the dynamic behavior of parameters.

Using the states in Σ, we construct *intervals* of time from Σ^+, the set of all nonempty, finite sequences of states. If s, t and u are states in Σ, then the following are possible intervals:

$\langle s \rangle$, $\langle sttsus \rangle$, $\langle tttt \rangle$.

Note that an interval always contains at least one state.

We now introduce some basic notation for manipulating intervals. Let us use I to denote the set of all intervals. For the moment, we take I to be the set Σ^+. Later on we will restrict I somewhat. Given an interval σ in I, we let $|\sigma|$ be the *length* of σ. Our convention is that an interval's length is the number of states *minus one*. Thus the intervals above have respective lengths 0, 5 and 3. The individual states of an interval σ are denoted by σ_0, σ_1, ..., $\sigma_{|\sigma|}$. For instance, the following equality is true iff the variable A has the value 5 in σ's final state:

$$\sigma_{|\sigma|}[\![A]\!] = 5.$$

The model described here views time as being discrete and is not intended to be a realistic representation of the world around us. Nonetheless, it provides a sound basis for reasoning about many interesting dynamic phenomena involving timing-dependent and functional behavior. Furthermore, a discrete-time view of the world often corresponds to our mental model of digital systems and computer programs. In any case, we can always make the granularity of time arbitrarily fine.

2.4 Interpretation of Expressions and Formulas

We now extend the interpretation M to give meaning to expressions and formulas on intervals. The construct $M_\sigma[\![e]\!]$ will be defined to equal the value in D of the expression e on the interval σ. Similarly, $M_\sigma[\![w]\!]$ will equal the truth value of the formula w on σ.

At first glance, the following definitions may seem somewhat arbitrary. We therefore suggest that an initial reading be rather cursory since the subsequent discussion and examples provide motivation. The definitions can then be referenced as needed.

- $M_\sigma[\![V]\!] = \sigma_0[\![V]\!]$, where V is a variable.
  Thus, a variable's value on an interval equals the variable's value in the interval's initial state.

- $M_\sigma[\![f(e_1, \ldots, e_k)]\!] = M[\![f]\!](M_\sigma[\![e_1]\!], \ldots, M_\sigma[\![e_k]\!])$.
  The interpretation of the function symbol f is applied to the interpretations of $e_1, \ldots, e_k$.

- $\mathcal{M}_\sigma[\![\bigcirc e]\!] = \mathcal{M}_{\langle\sigma_1...\sigma_{|\sigma|}\rangle}[\![e]\!]$, if $|\sigma| \geq 1$.
  We leave the value of $\bigcirc e$ unspecified on intervals having length 0.

- $\mathcal{M}_\sigma[\![p(e_1,\ldots,e_k)]\!] = \mathcal{M}[\![p]\!](\mathcal{M}_\sigma[\![e_1]\!],\ldots,\mathcal{M}_\sigma[\![e_k]\!])$.

- $\mathcal{M}_\sigma[\![e_1=e_2]\!] = true$ iff $\mathcal{M}_\sigma[\![e_1]\!] = \mathcal{M}_\sigma[\![e_2]\!]$.

- $\mathcal{M}_\sigma[\![\neg w]\!] = true$ iff $\mathcal{M}_\sigma[\![w]\!] = false$.

- $\mathcal{M}_\sigma[\![w_1 \wedge w_2]\!] = true$ iff
  $\mathcal{M}_\sigma[\![w_1]\!] = true$ and $\mathcal{M}_\sigma[\![w_2]\!] = true$.

- $\mathcal{M}_\sigma[\![\bigcirc w]\!] = true$ iff $|\sigma| \geq 1$ and $\mathcal{M}_{\langle\sigma_1...\sigma_{|\sigma|}\rangle}[\![w]\!] = true$.

- $\mathcal{M}_\sigma[\![\square w]\!] = true$ iff
  for all $i \leq |\sigma|$, $\mathcal{M}_{\langle\sigma_i...\sigma_{|\sigma|}\rangle}[\![w]\!] = true$.

Examples

We now illustrate the use of $\mathcal{M}$ by considering the semantics of the sample temporal formulas given earlier. Let s, t and u be states in which the variables I and J have the following values:

	I	J
s	1	2
t	3	4
u	3	1

The formula

$$(J = 2) \wedge \bigcirc(I = 3)$$

is true on an interval σ iff σ has length ≥ 1, the value of J in the state σ_0 is 2 and the value of I in the state σ_1 is 3. Thus, the formula is true on the interval $\langle stu \rangle$. On the other hand, the formula is false on the interval $\langle ttu \rangle$ because J's initial value on this interval is 4 instead of 2.

The formula

$$(\bigcirc \square[I = 3]) \wedge \neg([\bigcirc J] = 4)$$

is true on any interval σ having length ≥ 1 and in which I equals 3 in the states $\sigma_1, \ldots, \sigma_{|\sigma|}$ and J does not equal 4 in σ_1. Thus the formula is true on the interval $\langle sutut \rangle$ but is false on $\langle t \rangle$ and $\langle stutu \rangle$.

The formula

$$\bigcirc\bigl(\square[I = 3] \wedge \bigcirc\bigcirc[J = 4]\bigr)$$

is true on an interval σ having length ≥ 3 and in which the variable I equals 3 in the states $\sigma_1, \ldots, \sigma_{|\sigma|}$ and the variable J equals 4 in the state σ_3. Thus this formula is true of the interval $\langle suutu \rangle$ but is false on $\langle s \rangle$ and $\langle sutuu \rangle$.

2.5 Satisfiability and Validity

A formula w is *satisfied* by an interval σ iff the meaning of w on σ equals *true*:

$$\mathcal{M}_\sigma[\![w]\!] = true.$$

This is denoted as follows:

$$\sigma \models w.$$

If all intervals satisfy w then w is *valid*, written $\models w$.

Example (Validity):

The following formula is true on an interval σ iff $|\sigma| \geq 1$, the variable I always equals 1 and in the state σ_1, I equals 2:

$$\square(I = 1) \wedge \bigcirc(I = 2).$$

No interval can have all of these characteristics. Therefore the formula is false on all intervals and its negation is always true and hence valid:

$$\models \neg\bigl[\square(I = 1) \wedge \bigcirc(I = 2)\bigr].$$

3 Deriving Other Operators

The kinds of interval behavior one can describe with the constructs so far introduced may seem rather limited. In fact, this is not at all the case since we can develop quite a variety of derived operators. We now present a few that have proved useful in reasoning about simple computations.

3.1 Boolean Operators

The conventional boolean constructs $w_1 \vee w_2$ (*logical-or*), $w_1 \supset w_2$ (*implication*) and $w_1 \equiv w_2$ (*equivalence*) can be expressed in terms of $\neg$ and $\wedge$. We can define logical-or as shown below:

$$w_1 \vee w_2 \quad \equiv_{\text{def}} \quad \neg(\neg w_1 \wedge \neg w_2).$$

We then express implication and equivalence as follows:

$$w_1 \supset w_2 \quad \equiv_{\text{def}} \quad \neg w_1 \vee w_2,$$
$$w_1 \equiv w_2 \quad \equiv_{\text{def}} \quad (w_1 \supset w_2) \wedge (w_2 \supset w_1).$$

The boolean constructs *true* and *false* can also be derived as can the conditional formula

if w_1 then w_2 else w_3.

Example (Implication):

If in an interval σ, the variable I always equals 1 and in the state σ_1 the variable J equals 2 then it follows that the expression $I + J$ equals 3 in σ_1. This fact can be expressed by the following valid formula:

$$\models \quad [\Box(I = 1) \wedge \bigcirc(J = 2)] \supset \bigcirc(I + J = 3).$$

Example (Equivalence):
The formula

$$\bigcirc\big([I = 1] \wedge [J = 2]\big)$$

is true on an interval σ iff σ has length ≥ 1 and in the state σ_1, the variable I has the value 1 and the variable J has the value 2. It turns out that the conjunction

$$\bigcirc(I = 1) \wedge \bigcirc(J = 2)$$

has the same meaning. The equivalence of these two formulas is expressible as follows:

$$\bigcirc\big([I = 1] \wedge [J = 2]\big) \equiv \big[\bigcirc(I = 1) \wedge \bigcirc(J = 2)\big].$$

This formula is true on all intervals and is therefore valid. In general, if two formulas w_1 and w_2 have the same meaning on all intervals, then the equivalence $w_1 \equiv w_2$ is valid.

3.2 The Operator $\Diamond$

The construct $\Diamond w$ is true on an interval σ if there is some suffix subinterval on which the formula w is true:

$$\mathcal{M}_\sigma[\![\Diamond w]\!] = \mathit{true} \quad \text{iff} \quad \text{for some } i \leq |\sigma|,\ \mathcal{M}_{\langle\sigma_i\ldots\sigma_{|\sigma|}\rangle}[\![w]\!].$$

This behavior can be given in terms of the operators $\neg$ and $\Box$:

$$\Diamond w \quad \equiv_{\mathrm{def}} \quad \neg\,\Box\,\neg w.$$

Thus the operators $\Box$ and $\Diamond$ are in fact duals.

Example (Present and future):
The following formula illustrates important differences between various temporal constructs:

$$(I = 1) \wedge \bigcirc(I = 2) \wedge \Diamond(I = 3).$$

This is true on an interval σ having length at least 2 in which the variable I has the value 1 in the initial state σ_0, the value 2 in the next state σ_1 and eventually equals 3 in some subsequent state.

3.3 The Operators *empty* and *more*

The formula *empty* is true on an interval iff the interval has length 0:

$$\sigma \models empty \quad \text{iff} \quad |\sigma| = 0.$$

We can define *empty* as follows:

$$empty \quad \equiv_{\text{def}} \quad \neg \bigcirc true.$$

The formula *more* is true on an interval iff the interval has nonzero length. We can express *more* as follows:

$$more \quad \equiv_{\text{def}} \quad \bigcirc true.$$

From these definitions it readily follows that *more* is the opposite of *empty*:

$$\models \quad more \equiv \neg empty.$$

Example (Testing the length of an interval):

We can use the constructs $\bigcirc$ and *empty* to test the length of an interval. For example, the formula

$$\bigcirc \bigcirc \bigcirc empty$$

is true on an interval σ iff σ has length 3.

3.4 The Operators *gets* and *stable*

It is sometimes necessary to say that over time one expression e_1 equals another expression e_2 but with a one-unit delay. We use the construct e_1 *gets* e_2 to represent this and define it as follows:

$$e_1 \; gets \; e_2 \quad \equiv_{\text{def}} \quad \square\big(more \supset [(\bigcirc e_1) = e_2]\big).$$

The test *more* ensures that we do not "run off" the edge of the interval by erroneously attempting to examine the value of the expression e_1 in the nonexistent state $\sigma_{|\sigma|+1}$.

For instance, the formula K *gets* $2K$ is true on an interval σ iff the variable K is repeatedly doubled from each state to its successor:

$$\sigma \models \quad K \; gets \; 2K \quad \text{iff} \quad \text{for all } i < |\sigma|,\; \sigma_{i+1}[\![K]\!] = 2 \cdot \sigma_i[\![K]\!].$$

The construct *stable e* is true iff the value of the expression e remains unchanged. We can readily define *stable* in terms of *gets*:

stable e $\equiv_{\text{def}}$ *e gets e*.

Example (Expressing an invariant condition):
The following formula is true on an interval σ in which I and J are both initially 0 and I repeatedly increases by 1 and J repeatedly increases by 2:

$$(I = 0) \wedge (J = 0) \wedge (I \text{ gets } I + 1) \wedge (J \text{ gets } J + 2).$$

In any interval for which this is true, J always equals $2I$. Below is a valid property that formalizes this:

$$\models \quad [(I=0) \wedge (J=0) \wedge (I \text{ gets } I+1) \wedge (J \text{ gets } J+2)] \supset \Box(J=2I).$$

This shows how the operator $\Box$ can express an invariant condition.

Example (Stability):
The formula

$$(I = 1) \wedge \text{stable } I$$

is true iff I initially equals 1 and its value remains unchanged. This is the same as saying that I always equals 1. The following valid property expresses this equivalence:

$$\models \quad [(I=1) \wedge \text{stable } I] \equiv \Box(I=1).$$

3.5 The Operator *halt*

We can specify that a formula w becomes true only at the end of an interval σ by using the formula *halt w*:

halt w $\equiv_{\text{def}}$ $\Box(w \equiv \text{empty})$.

Thus w must be false until the last state at which time w is true. For example, the formula

halt$(I > 100)$

is true on σ iff the value of the variable I exceeds 100 in exactly the last state of σ.

Example (Repeatedly doubling a number):

From what we have so far presented, it can be seen that the formula

$$(I = 1) \land halt(I > 100) \land (I \text{ gets } 2I)$$

is true on an interval where the variable I is initially 1 and repeatedly doubles until it exceeds 100. The following valid implication states that intervals on which this formula is true will terminate upon I equalling the value 128:

$$\models \quad [(I = 1) \land halt(I > 100) \land (I \text{ gets } 2I)] \supset halt(I = 128).$$

3.6 Temporal Equality

The construct $e_1 \approx e_2$ is called *temporal equality* and is true iff the expressions e_1 and e_2 are always equal:

$$e_1 \approx e_2 \quad \equiv_{\text{def}} \quad \Box(e_1 = e_2).$$

Example (Computing factorials):

Consider the following formula for running through factorials:

$$(I = 0) \land (I \text{ gets } I + 1) \land (J \approx I!).$$

The value of J can be seen to start at 1 and then repeatedly be multiplied by $I + 1$. This is expressed by the following property:

$$\models \quad [(I = 0) \land (I \text{ gets } I + 1) \land (J \approx I!)] \supset$$
$$[(J = 1) \land (J \text{ gets } [I + 1] \cdot J)].$$

3.7 The Operator *weak next*

In order for the construct $\bigcirc w$ to be true on an interval σ, the length of σ must be at least 1. We therefore refer to this as *strong next*. The related construct $\circledcirc w$ is called *weak next* and is true on an interval σ if either σ has length 0 or the subformula w is true on $\langle \sigma_1 \ldots \sigma_{|\sigma|} \rangle$. We can express *weak next* in terms of *strong next*:

$$\circledcirc w \quad \equiv_{\text{def}} \quad empty \lor \bigcirc w.$$

The operator *weak next* provides a concise and natural way to express a construct as a conjunction of its immediate effect and future effect. Here are some examples:

$$\begin{aligned}
\models\ &\square\, w &\equiv\ & w \wedge \textcircled{w}\, \square\, w, \\
\models\ &\bigcirc w &\equiv\ & \mathit{more} \wedge \textcircled{w}\, w, \\
\models\ &e_1\ \mathit{gets}\ e_2 &\equiv\ & \bigl[\mathit{more} \supset ([\bigcirc e_1] = e_2)\bigr] \wedge \textcircled{w}(e_1\ \mathit{gets}\ e_2), \\
\models\ &\mathit{halt}\ w &\equiv\ & (\mathit{empty} \equiv w) \wedge \textcircled{w}(\mathit{halt}\ w), \\
\models\ &e_1 \approx e_2 &\equiv\ & (e_1 = e_2) \wedge \textcircled{w}(e_1 \approx e_2).
\end{aligned}$$

These kinds of equivalences turn out to be useful in the design of interpreters.

4 Programming in Temporal Logic

Consider the formula

$$(M = 4) \land (N = 1) \\ \land \, halt(M = 0) \land (M \text{ gets } M - 1) \land (N \text{ gets } 2N). \quad (4.1)$$

This holds true of intervals of length 4 in which M successively runs through the values 4, 3, 2, 1 and 0 and N simultaneously runs through the values 1, 2, 4, 8, and 16. Let us now explore how to automate the process of taking such a temporal formula and finding an interval satisfying it. One way to do this is by developing a procedure that analyzes the formula and determines the behavior of all free variables in every state. The result can itself be expressed as a temporal formula. For instance, here is one way to represent the result for formula (4.1):

$$\big([M = 4] \land [N = 1]\big) \\ \land \bigcirc \big([M = 3] \land [N = 2]\big) \\ \land \bigcirc \bigcirc \big([M = 2] \land [N = 4]\big) \\ \land \bigcirc \bigcirc \bigcirc \big([M = 1] \land [N = 8]\big) \\ \land \bigcirc \bigcirc \bigcirc \bigcirc \big([M = 0] \land [N = 16] \land \text{empty}\big).$$

Note that this formula is logically equivalent to the original formula (4.1). We can view it as a kind of normal form containing state-by-state behavior of all free variables. The process of determining such a normal form can be computerized. It is in essence a form of program execution where the original formula represents the program and the resulting normal form corresponds to the actual computation.

The general problem of finding a normal form for an arbitrary temporal formula is unsolvable. However, there are subsets of temporal logic for which the task is manageable. We have developed *Tempura*, a programming language based on a subset that seems by experience to be efficiently implementable and of use in describing interesting and practical computations. For example, formula (4.2) is a legal Tempura program which when executed produces the output shown in figure 4.1.

$$(M = 4) \wedge (N = 1) \wedge halt(M = 0) \\ \wedge (M \ gets \ M - 1) \wedge (N \ gets \ 2N) \wedge \Box \ display(M, N). \qquad (4.2)$$

This repeatedly prints the values of M and N by means of the *display* construct. Note that the program's behavior is unaffected even if we change the order of the conjunction's operands. For instance, the following variant reverses them:

$$\Box \ display(M, N) \wedge (N \ gets \ 2N) \wedge (M \ gets \ M - 1) \\ \wedge halt(M = 0) \wedge (N = 1) \wedge (M = 4).$$

During the execution of the following program, the user is continually asked for the values of I by means of the *request* construct:

$$\Box \ request(I) \wedge halt(I = 0) \wedge (J = 0) \\ \wedge (J \ gets \ J + I) \wedge \Box \ display(J). \qquad (4.3)$$

These values are summed into J and J itself is displayed. The interval terminates upon I equalling 0. A typical session is given in figure 4.2. Numbers in boxes (e.g., $\boxed{6}$) are input by the user.

Figure 4.1: Execution of formula (4.2)

```
State  0:  M= 4    N= 1
State  1:  M= 3    N= 2
State  2:  M= 2    N= 4
State  3:  M= 1    N= 8
State  4:  M= 0    N=16

Done!  Computation length = 4.
```

4.1 Syntax of Tempura

Let us now look at the basic syntax of Tempura. In later chapters, as new temporal logic constructs are introduced, variants of them will be added to Tempura. The main syntactic categories in Tempura are locations, expressions and statements.

4.1.1 Locations

A *location* is a place where values are stored and examined. Variables such as I, J and K are permissible locations. In addition, if l is a location, so is the temporal construct $\bigcirc l$.

4.1.2 Expressions

Expressions can be either arithmetic or boolean. All numeric constants and variables are legal arithmetic expressions. In addition, if e_1 and e_2 are arithmetic expressions, so are the following operations:

$$e_1 + e_2, \quad e_1 - e_2, \quad e_1 \cdot e_2, \quad e_1 \div e_2, \quad e_1 \bmod e_2.$$

Relations such as $e_1 = e_2$ and $e_1 \geq e_2$ are boolean expressions. If b, b_1 and b_2 are boolean expressions, then so are the following:

$$\neg b, \quad b_1 \wedge b_2, \quad b_1 \vee b_2, \quad b_1 \supset b_2, \quad b_1 \equiv b_2.$$

The constants *true* and *false* and the temporal constructs *empty*

Figure 4.2: Execution of formula (4.3)

```
State 0: I = 6
State 0: J = 0
State 1: I = 2
State 1: J = 6
State 2: I = 5
State 2: J = 8
State 3: I = 0
State 3: J = 13

Done! Computation length = 3.
```

and *more* are boolean expressions as well. In addition, the conditional expression

$$\textit{if } b \textit{ then } e_1 \textit{ else } e_2$$

is permitted. Here e_1 and e_2 can be either arithmetic or boolean expressions.

4.1.3 Statements

Certain temporal formulas are legal statements in Tempura. A statement is either *simple* or *compound*. Simple statements are built from the constructs given below:

true	(no-operation)
false	(abort)
$l = e$	(simple assignment)
empty	(terminate)
more	(do not terminate).

The statement $l = e$ stores the value of the arithmetic expression e in the location l. In addition to these statements, the following can be used for requesting and displaying values:

$request(l_1, \ldots, l_n)$	(request values of locations)
$display(e_1, \ldots, e_n)$	(display values of expressions).

Compound statements are built from the constructs given below. Here w, w_1 and w_2 are themselves statements and b is a boolean expression:

$w_1 \wedge w_2$	(parallel composition)
$b \supset w$	(implication)
$\ocircle \, w$	(weak next)
$\square \, w$	(always)

Note that certain temporal formulas can be used as both boolean expressions and statements. Here are three examples:

$$I = 3, \quad (J = 2) \wedge (K = J + 3), \quad (I = 0) \supset \textit{empty}.$$

On the other hand, the following legal boolean expressions are not Tempura statements even though they are semantically equivalent to the respective formulas given above:

$$3 = I, \quad (2 = J) \wedge (J + 3 = K), \quad \neg(I = 0) \vee \textit{empty}.$$

4.2 Some Other Statements

Other constructs such as *gets*, *stable* and *halt* can be readily added to Tempura. One way to do this is to expand these to statements already described. Here are some possible equivalences:

$$\begin{aligned}
\textit{if } b \textit{ then } w_1 \textit{ else } w_2 &\equiv (b \supset w_1) \wedge (\neg b \supset w_2), \\
\bigcirc w &\equiv \textit{more} \wedge \textcircled{w} \, w, \\
l \textit{ gets } e &\equiv \Box\bigl(\neg \textit{empty} \supset [(\bigcirc l) = e]\bigr), \\
\textit{stable } l &\equiv l \textit{ gets } l, \\
\textit{halt } b &\equiv \Box(\textit{if } b \textit{ then empty else more}), \\
l \approx e &\equiv \Box(l = e).
\end{aligned}$$

An alternative approach is to include these features directly in the base language.

4.3 Determinism

As we mentioned earlier, Tempura statements are limited to a subset of temporal formulas. However, even syntactically legal programs may possibly be nonexecutable. This is because the interpreter expects the user to completely specify the behavior of program variables and to indicate when termination should occur. For example, the formula

$$I \textit{ gets } (I + 1)$$

lacks information on I's initial value and does not specify when to stop. Thus it cannot by itself be transformed to any particular computation sequence on I and is therefore not considered a complete program. Other details must be included for the interpreter to operate properly. For similar reasons compound statements built using the operators $\vee$ and $\Diamond$ are not permitted. Of course, we could be more lenient by using backtracking and related techniques to resolve any ambiguities. However, for the sake of the simplicity and efficiency of the interpreter, it seems reasonable at the moment to require explicit information on all aspects of variable behavior.

5 Some Additional Constructs

Let us now consider how to add three important features to the temporal logic. These are *static variables*, *lists* and *quantifiers*. We subsequently use them as a basis for deriving one operator that specifies interval length and another that describes in-place assignment.

5.1 Static Variables

For a given variable A and an interval σ, it is possible for A to have a different value in each of σ's states $\sigma_0, \sigma_1, \ldots, \sigma_{|\sigma|}$. For this reason, A is called a *state variable*. It turns out to be useful to introduce a category of variables called *static variables*. Our convention is to have them start with a lower-case letter (e.g., a, *tree* and aBc). We now constrain the set $\mathcal{I}$ of permitted intervals so that for any interval σ in $\mathcal{I}$ and any static variable a, the values of a on σ's states are all identical:

$$\mathcal{M}_{\sigma_0}[\![a]\!] = \mathcal{M}_{\sigma_1}[\![a]\!] = \cdots = \mathcal{M}_{\sigma_{|\sigma|}}[\![a]\!].$$

Thus a static variable is stable:

$$\models \quad stable\ a.$$

Identifiers starting with upper-case letters (e.g., A and *Tree*) remain state variables and can change from state to state. Furthermore, even a static variable can have different values on two distinct intervals σ and σ'.

Example (Computing powers):
The following formula has the variable J successively equal the powers m^0, m^1, m^2, ..., m^n:

$$(I = 0) \land (J = 1) \land (I \text{ gets } I+1) \land (J \text{ gets } m \cdot J) \land \text{halt}(I = n).$$

The variables m and n are static and therefore do not need to be kept explicitly stable.

5.2 Quantified Formulas

We permit formulas of the form

$$\exists V : w,$$

where V is any variable and w is itself a formula. This is called *existential quantification*. Note that V can be either a static or state variable. Below is an instance of this construct:

$$\exists I : \Box (J = 2I).$$

Existential quantification readily generalizes to many variables:

$$\exists V_1, V_2, \ldots, V_n : w \quad \equiv \quad \exists V_1 : (\exists V_2 : (\ldots (\exists V_n : w))).$$

Universal quantification has the form $\forall V : w$ and is defined as the dual of existential quantification:

$$\forall V : w \quad \equiv_{\text{def}} \quad \neg \exists V : \neg w.$$

Here are the semantics of $\exists$:

$$\mathcal{M}_\sigma [\![\exists V : w]\!] = \textit{true} \quad \text{iff}$$
$$\quad \text{for some interval } \sigma' \in \mathcal{I},\ \sigma \sim_V \sigma' \text{ and } \mathcal{M}_{\sigma'}[\![w]\!] = \textit{true}.$$

The relation $\sigma \sim_V \sigma'$ is defined to be true iff the intervals σ and σ' have the same length and agree on the behavior of all variables except possibly the variable V.

Example

Consider the following states and their assignments to the variables I and J:

	I	J
s	2	4
t	0	4
u	2	3

We assume that s, t and u agree on assignments to all other variables.

The formula

$$\exists I. \ \Box(J = 2I)$$

is intuitively true on any interval on which we can construct an I such that J always equals $2I$. This is the same as saying that J is always even. For example, the interval $\langle ttt \rangle$ satisfies the formula. From the semantics of $\exists$ given previously it follows that to show this we need to construct an interval σ' for which the relation $\langle ttt \rangle \sim_I \sigma'$ is true and which satisfies the subformula $\Box(J = 2I)$. The interval $\langle sss \rangle$ achieves both of these constraints. Therefore $\langle ttt \rangle$ satisfies the original formula. Other intervals satisfying the formula include $\langle sss \rangle$ itself and $\langle sst \rangle$ but not $\langle u \rangle$ or $\langle stut \rangle$. Existential quantification is a tricky concept and the reader should not necessarily expect to grasp it immediately.

Example (Hiding a variable):

The formula below has J always equalling twice the value of a hidden variable I. The value of I is initially 0 and repeatedly increases by 1:

$$\exists I \colon \big[(I = 0) \land (I \text{ gets } I + 1) \land (J \approx 2I)\big].$$

This is logically equivalent to initializing J to 0 and repeatedly increasing it by 2:

$$(J = 0) \land (J \text{ gets } J + 2).$$

We can express this equivalence as the following property:

$$\begin{aligned}\models \ &\big(\exists I \colon \big[(I = 0) \land (I \text{ gets } I + 1) \land (J \approx 2I)\big]\big) \\ &\equiv \big[(J = 0) \land (J \text{ gets } J + 2)\big].\end{aligned}$$

The following formula has two distinct variables that are both called I:

$$(I = 0) \land (I\ gets\ I+1) \land halt(I = 5) \land \exists I\colon \bigl[(I = 1) \land (I\ gets\ 3I)\bigr].$$

The first I runs from 0 to 5 and in parallel the second I is repeatedly tripled from 1 to 243. The use of existential quantification ($\exists$) keeps the two I's separate and in effect hides the second one. In fact the formula is logically equivalent to its subformula

$$(I = 0) \land (I\ gets\ I + 1) \land halt(I = 5).$$

As these examples illustrate, the operator $\exists$ provides a means of creating locally scoped variables.

5.3 Enlarging the Data Domain

Until now we have assumed that the underlying data domain $\mathcal{D}$ consists of the integers. Let us now enlarge it to include the boolean values *true* and *false* as well as nested finite lists of values. Here are some sample lists:

$$\langle 3 \rangle, \quad \langle\rangle, \quad \langle \langle true, 2\rangle, \langle\rangle, \langle 1, \langle 2, false\rangle\rangle \rangle.$$

The following constructs are now included among the temporal logic's expressions and are later added to Tempura:

- Simple list construction: $\langle e_0, \ldots, e_{n-1}\rangle$,
  where $n \geq 0$ and $e_0, \ldots, e_{n-1}$ are expressions.

- Iterative list construction: $\langle e_1\colon v < e_2\rangle$,
  where e_1 and e_2 are expressions and v is a static variable.

- Subscripting: $e_1[e_2]$,   where e_1 and e_2 are expressions.

- List length: $|e|$,   where e is an expression.

In addition, there is the following new predicate:

- List predicate: $list(e_1, e_2)$,   where e_1 and e_2 are expressions.

The semantics of the various expressions for manipulating lists are as one might expect. Our convention regarding subscripting is to index from the left starting with 0. For example, the following expressions are all equal:

$$3, \quad |\langle \mathit{false}, 1, 4\rangle|, \quad \langle 0, \mathit{true}, 3\rangle[2], \quad \langle 7 - i : i < 8\rangle[4].$$

The predicate $\mathit{list}(e_1, e_2)$ is true on an interval if the initial value of the expression e_1 is a list of length e_2.

When no ambiguity arises, we express a subscripted expression such as $L[i]$ using the notation L_i.

Example (Computing lists of powers):

Here is one way to compute lists of successive powers of 0, 1, 2 and 3:

$$(I = 0) \wedge (I \textit{ gets } I + 1) \wedge \mathit{halt}(I = 10) \wedge \Box(L = \langle 0^I, 1^I, 2^I, 3^I\rangle).$$

Alternatively, the following formula can be used:

$$(I = 0) \wedge (I \textit{ gets } I + 1) \wedge \mathit{halt}(I = 10) \wedge [\Box \, \mathit{list}(L, 4)]$$
$$\wedge (L_0 \approx 0^I) \wedge (L_1 \approx 1^I) \wedge (L_2 \approx 2^I) \wedge (L_3 \approx 3^I).$$

We now present a number of useful operators derived from the constructs so far introduced.

5.4 The Operator *len*

The formula $\mathit{len}(e)$ is true on an interval having length exactly e:

$$\mathcal{M}_\sigma[\![\mathit{len}(e)]\!] = \mathit{true} \quad \text{iff} \quad \mathcal{M}_\sigma[\![e]\!] = |\sigma|.$$

It turns out that we can express *len* by means of existential quantification and other previously introduced constructs. Here is one way:

$$\mathit{len}(e) \quad \equiv_{\mathrm{def}} \quad \exists I : \bigl[(I = e) \wedge (I \textit{ gets } I - 1) \wedge \mathit{halt}(I = 0)\bigr],$$

where I does not occur freely in e. We use I as a hidden counter that is initialized to e's value and then keeps track of how much time remains in the interval.

Example (Doubling a variable):

The following formula has the variable N run through the first few powers of 2:

$$len(5) \wedge (N = 1) \wedge (N \text{ gets } 2N).$$

The *len* construct is used to specify the length of the computation.

5.5 Bounded Quantification

We now introduce formulas of the form

$$\forall v < e: w,$$

where v is a static variable, e is an expression and w is a formula. The construct is referred to as *bounded universal quantification* and can be defined as follows:

$$\forall v < e: w \quad \equiv_{\text{def}} \quad \forall v: (0 \leq v < e \supset w).$$

This is especially useful for processing elements of a list variable in parallel. Bounded existential quantification can be defined in an analogous way. In addition, it is easy to generalize this notation to handle ranges such as $v \leq e$ and $e_1 \leq v < e_2$.

Example (Generalized computation of powers):

The following formula successively assigns the list L the first $n+1$ powers of the numbers $0, 1, \ldots, m-1$:

$$\exists I: \big[(I = 0) \wedge (I \text{ gets } I - 1) \\ \wedge \, halt(I = n) \wedge (L \approx \langle k^I: k < m \rangle)\big]. \tag{5.1}$$

Figure 5.1 shows the behavior of L for $m = 5$ and $n = 6$. The actual program consists of the conjunction of the above formula with the statement

$$\square \; display(L).$$

Here is another technique which is logically equivalent and uses bounded universal quantification to simultaneously manipulate each of the elements of L:

$$len(n) \wedge [\square \; list(L, m)] \\ \wedge \, \forall k < m: \big[(L_k = 1) \wedge (L_k \text{ gets } k \cdot L_k)\big].$$

5.6 The Operator *fin*

The formula *fin w* is true on an interval σ iff the formula *w* is itself true on the final subinterval $\langle \sigma_{|\sigma|} \rangle$. We express *fin w* as follows:

$$\textit{fin } w \quad \equiv_{\text{def}} \quad \Box(\textit{empty} \supset w).$$

The formula *fin w* is weaker than *halt w* since *fin w* only looks at the last state whereas *halt w* tests behavior throughout.

Example (Doubling a variable):

The following formula is true on an interval σ iff $|\sigma| = 3$ and I is initially 1 and repeatedly doubles:

$$len(3) \wedge (I = 1) \wedge (I \textit{ gets } 2I).$$

One effect is that I ends up equal to 8. This is expressed by the valid implication given below:

$$\models \quad \bigl[len(3) \wedge (I = 1) \wedge (I \textit{ gets } 2I)\bigr] \supset \textit{fin}(I = 8).$$

5.7 Temporal Assignment

The formula $e_1 \to e_2$ is true for an interval if the initial value of the expression e_1 equals the final value of the expression e_2. We define this as follows:

$$e_1 \to e_2 \quad \equiv_{\text{def}} \quad \exists a \colon \bigl[(a = e_1) \wedge \textit{fin}(e_2 = a)\bigr],$$

Figure 5.1: Execution of formula (5.1)

```
State  0:  L=<1,1,1,1,1>
State  1:  L=<0,1,2,3,4>
State  2:  L=<0,1,4,9,16>
State  3:  L=<0,1,8,27,64>
State  4:  L=<0,1,16,81,256>
State  5:  L=<0,1,32,243,1024>
State  6:  L=<0,1,64,729,4096>

Done!  Computation length = 6.
```

where the static variable a does not occur freely in either e_1 or e_2. The stability of the value of a is used to compare the values of e_1 and e_2 at different times. We call this construct *temporal assignment*. For example, the formula $I + 1 \to I$ is true on an interval σ iff the value of $I + 1$ in the initial state σ_0 equals the value of I in the final state $\sigma_{|\sigma|}$. If desired, we can reverse the direction of the arrow:

$$I \gets I + 1.$$

The formula

$$(I \gets I + 1) \wedge (J \gets J + I)$$

is then true on an interval iff I increases by 1 and in parallel J increases by I. Similarly, the following specifies that the values of the state variables A and B are exchanged:

$$(A \to B) \wedge (B \to A).$$

Unlike the assignment statement in conventional programming languages, temporal assignment only affects variables explicitly mentioned; the values of other variables do not necessarily remain fixed. For example, the formulas

$$I \gets I + 1$$

and

$$(I \gets I + 1) \wedge (J \gets J)$$

are not equivalent since the first formula does not require J's initial and final values to be equal. Thus, temporal assignment lacks the so-called *frame assumption*.

Example (Maximum of two numbers):
The temporal formula

$$\textit{if } I \geq J \textit{ then } (I \gets I) \textit{ else } (I \gets J)$$

is true in any interval where I's value in the final state equals the maximum of the values of I and J in the initial state. This can be seen by case analysis on the test $I \geq J$.

Let the function $max(i,j)$ equal the maximum of the two values i and j. The following temporal formula therefore places the maximum of I and J into I:

$$I \leftarrow max(I,J)$$

The equivalence of the two approaches is expressed by the following property:

$$\models \quad [I \leftarrow max(I,J)] \equiv [\textit{if } I \geq J \textit{ then } (I \leftarrow I) \textit{ else } (I \leftarrow J)].$$

5.8 Incorporating these Constructs into Tempura

We now extend Tempura to allow locations, expressions and statements in the ways described below.

5.8.1 Locations

Static variables such as a and x are now permitted as locations. In addition if l is a location and e is an expression, then the subscript construct $l[e]$ is a permitted location.

5.8.2 Expressions

The following are now legal expressions:

v	(static variables)
$\langle e_0, \ldots, e_{n-1} \rangle$	(simple list constructor)
$\langle e_1 : v < e_2 \rangle$	(iterative list constructor)
$e_1[e_2]$	(subscripting)
$\|e\|$	(list length).

Here v is a static variable and e, e_0, e_1, ..., e_{n-1} are all expressions.

In addition, we permit function invocations of the form

$$f(e_1, \ldots, e_n)$$

where f is a function defined in the manner described later and $n \geq 0$ and e_1, ..., e_n are expressions.

5.8.3 Statements

We now allow variables to be assigned not only integers, but boolean and list values are well. Here are some examples:

$$Done = true, \quad L = \langle 1, 2, 4 \rangle, \quad \textit{Flag gets } (\neg \textit{Flag}).$$

In addition, the following are permitted statements:

$list(l, e)$	(list declaration)
$len(e)$	(interval length)
$\exists V_1, \ldots, V_n : w$	(existential quantification)
$\forall v < e : w$	(bounded universal quantification)
$fin\ w$	(terminal statement)
$l \leftarrow e$	(temporal assignment).

Here l is a location, e is an expression, V is any variable, v is a static variable, and w is a statement.

We also permit predicate invocations of the form

$$p(e_1, \ldots, e_n)$$

where p is a predicate defined in the manner described below and $n \geq 0$ and $e_1, \ldots, e_n$ are expressions. The interpreter uses call-by-reference when passing parameters.

Statements of the following forms can be used to define functions and predicates respectively:

$$\textit{define } f(V_1, \ldots, V_n) = e,$$
$$\textit{define } p(V_1, \ldots, V_n) \equiv w.$$

The formal parameters $V_1, \ldots, V_n$ are state variables or static variables and e is an expression and w is a statement. The identifier used in place of f or p should be static. Here are two sample definitions:

$$\textit{define } min(i, j) = (\textit{if } i \leq j \textit{ then } i \textit{ else } j),$$
$$\textit{define } double(M) \equiv (M \textit{ gets } 2M).$$

Note that recursive definitions are permitted. Furthermore, the body of a definition can include temporal constructs. Thus our actual usage of predicates and functions is more general than indicated in the temporal logic semantics presented earlier in section 2.3. It is not difficult to adjust the semantics to take this into account although we omit the details.

6 The Operator *chop*

Temporal logic contains various constructs such as *chop* and *while* that are rather similar to certain kinds of statements found in Algol and related programming languages. We first extend the syntax and semantics of the temporal logic to include *chop*. The resulting formalism is called Interval Temporal Logic. Within it we define a number of interval-dependent operators and subsequently expand Tempura to include them.

6.1 Syntax and Semantics of *chop*

We now permit formulas of the following form:

$w_1; w_2$, where w_1 and w_2 are formulas.

The operator ";" is known as *chop*. A formula $w_1; w_2$ is true on an interval σ iff there is at least one way to split σ into two subintervals such that the formula w_1 is true on the left subinterval and the formula w_2 is true on the right subinterval:

$$\mathcal{M}_\sigma[\![w_1; w_2]\!] = true \quad \text{iff}$$
$$\text{for some } i \leq |\sigma|,$$
$$\mathcal{M}_{\langle\sigma_0...\sigma_i\rangle}[\![w_1]\!] = true \text{ and } \mathcal{M}_{\langle\sigma_i...\sigma_{|\sigma|}\rangle}[\![w_2]\!] = true.$$

Note that the two subintervals $\langle\sigma_0 \ldots \sigma_i\rangle$ and $\langle\sigma_i \ldots \sigma_{|\sigma|}\rangle$ share the state σ_i.

Example (Sequential composition of assignments):
The formula

$$(K + 1 \to K); (K + 2 \to K)$$

is true on an interval σ iff there is some $i \leq |\sigma|$ such that the subformula $K+1 \to K$ is true on the subinterval $\langle \sigma_0 \ldots \sigma_i \rangle$ and the subformula $K+2 \to K$ is true on the remaining subinterval $\langle \sigma_i \ldots \sigma_{|\sigma|} \rangle$. The net effect is that K increases by 3. This is expressed by the following property:

$$\models \quad [(K+2 \to K); (K+1 \to K)] \supset (K+3 \to K).$$

Example (Using *chop* to express $\Diamond$ and $\Box$):

By varying the operands of *chop*, we can selectively examine different kinds of subintervals. For example, a formula of the form

true; *w*

is true on an interval if the formula w is true on some suffix subinterval. Thus this provides a way of expressing the operator $\Diamond$ and consequently $\Box$ as its dual. Similarly, a formula of the form

true; *w*; *true*

is true on an interval if w is true on some arbitrary subinterval. Since *chop* is associative, we can omit parentheses without being ambiguous.

6.2 Discussion of the Operator *chop*

The construct *chop* is rather different from the conventional temporal operators $\Box$ and $\bigcirc$. The latter examine an interval's suffix subintervals whereas *chop* splits the interval and tests both parts. This facilitates looking at arbitrary subintervals of time.

Harel, Kozen and Parikh [16] appear to be the first to mention *chop* as a temporal construct. It is considered in more detail by Chandra, Halpern, Meyer and Parikh [6]. In references [14] and [34] we use *chop* to facilitate reasoning about time-dependent digital hardware. Our subsequent work in [37] and [35] uses *chop* to give specifications and properties of simple algorithms and message-passing systems. In the rest of this section we examine *chop* and other ITL constructs and then extend Tempura to include some of them.

6.3 Simple For-Loops

The following simple for-loop repeats w for e times in succession, where w is a formula and e is an expression:

for e times do w.

This can also be written w^e. Here is one way to state that the variable I increases by 1 for 4 times in succession:

for 4 times do $(I + 1 \to I)$.

This is equivalent to the formula

$$(I + 1 \to I); (I + 1 \to I); (I + 1 \to I); (I + 1 \to I).$$

In the case of zero iterations, the for-loop is equivalent to *empty*:

$$\models \quad (\textit{for } 0 \textit{ times do } w) \equiv \textit{empty}.$$

Note that there is no requirement that the iterations of a loop take any time. Thus, a formula such as

for 5 times do $(I \to I)$

is readily satisfied by an empty interval:

$$\models \quad \textit{empty} \supset \bigl[\textit{for } 5 \textit{ times do } (I \to I)\bigr].$$

The formula w^* (read "w *star*") is true if the subformula w occurs some number of times in succession. This operator is sometimes known as *chop-star*. We can express it in terms of a simple for-loop:

$$w^* \quad \equiv_{\text{def}} \quad \exists n \colon (n \geq 0 \wedge [\textit{for } n \textit{ times do } w]),$$

where n does not occur freely in w. For example, the formula

$$(I + 1 \to I)^*$$

is true on an interval if I repeatedly increases over some unspecified number of iterations.

6.4 Indexed For-Loops

In addition to the simple for-loop, we permit an indexed variant having the following syntax:

$for\ v < e\ do\ w.$

Here v is a static variable, e is an expression and w is a formula. For instance, the formula

$for\ k < 4\ do\ (I + k \to I).$

is equivalent to

$$(I + 0 \to I); (I + 1 \to I); (I + 2 \to I); (I + 3 \to I).$$

6.5 While-Loops and Related Constructs

The *temporal while-loop* is another important construct in ITL. The basic form is similar to that of a while-loop in Algol:

$while\ w_1\ do\ w_2.$

Both w_1 and w_2 are themselves formulas. The while-loop obeys the following general expansion property:

$while\ w_1\ do\ w_2\ \equiv$
$\quad if\ w_1\ then\ \bigl(w_2; [while\ w_1\ do\ w_2]\bigr)\ else\ empty.$

Thus, if w_1 is true, the body of the loop, w_2, is examined after which the loop is repeated. If w_1 is false, the interval must have length 0. The *chop-star* operator can in fact be derived from a while-loop:

$w^* \quad \equiv \quad while\ more\ do\ w.$

A *repeat-loop* has the form

$repeat\ w_1\ until\ w_2$

and can be expressed using a while-loop:

$repeat\ w_1\ until\ w_2 \quad \equiv_{def} \quad w_1; (while\ \neg w_2\ do\ w_1).$

Another loop construct has the form

$$loop\ w_1\ exit\ when\ w_2\ otherwise\ w_3,$$

where w_1, w_2 and w_3 are themselves formulas. The formula w_2 is used to exit from the loop. Here is how to express the loop in terms of the *while* construct:

$$loop\ w_1\ exit\ when\ w_2\ otherwise\ w_3 \quad \equiv_{\text{def}}$$
$$w_1; (while\ \neg w_2\ do\ [w_3; w_1]).$$

In section 8.2 we use this style of loop to describe the basic algorithm used by a Tempura interpreter.

Example (Greatest common divisor):

Consider the following assignment which specifies that N's final value equals the initial value of the greatest common divisor of M and N:

$$N \leftarrow gcd(M, N).$$

The formula below implies this:

$$while\ (M \neq 0)\ do\ \bigl([M \leftarrow N\ mod\ M] \wedge [N \leftarrow M]\bigr).$$

6.6 Deriving For-Loops and While-Loops

Let us look at one way to express for-loops and while-loops entirely in terms of *chop* and other ITL constructs already introduced. This discussion can be omitted by the reader. We first define the predicate $end\_points(l, n)$ to be true on intervals where l is a list of $n + 1$ offsets indicating the end points of some successive subintervals corresponding to iterations:

$$end\_points(l, n) \quad \equiv_{\text{def}}$$
$$list(l, n + 1) \wedge (l_0 = 0) \wedge len(l_n) \wedge \forall i < n: (l_i \leq l_{i+1}).$$

Thus the element l_0 equals 0 since this is the start of any initial iteration. The final element l_n analogously equals the length of the interval. Furthermore, the sequence formed by l's elements is weakly increasing.

We also make use of the construct *iteration*(l, n, i, w). This is true on an interval if the formula w is true on the subinterval corresponding to the i-th iteration with respect to the values of l and n (i.e., the subinterval bound by l_i and l_{i+1}). Here is one way to express *iteration*:

$$iteration(l, n, i, w) \equiv_{\text{def}} len(l_i); w; len(l_n - l_{i+1}).$$

The simple for-loop construct can now be expressed as follows:

for e times do w $\equiv_{\text{def}}$
$\exists l, n\colon \big([n = e]$
  $\land\ end\_points(l, n)$
  $\land\ \forall i < n\colon iteration(l, n, i, w)$
$\big),$

where i, l and n do not occur freely in e or w. The technique for the indexed for-loop is analogous:

for v < e do w $\equiv_{\text{def}}$
$\exists l, n\colon \big([n = e]$
  $\land\ end\_points(l, n)$
  $\land\ \forall i < n\colon iteration(l, n, i, \exists v\colon [v = i \land w])$
$\big),$

where i, l and n do not occur freely in v, e or w.

Below is the derivation of the while-loop:

while w_1 do w_2 $\equiv_{\text{def}}$
$\exists l, n\colon \big(end\_points(l, n)$
  $\land\ \forall j < n+1\colon \big[(j < n) \equiv \big(len(l_j); w_1\big)\big]$
  $\land\ \forall i < n\colon iteration(l, n, i, w_2)$
$\big).$

Here the formula w_1 must be true at the beginning of every iteration but false at the end. Furthermore, the formula w_2 must be true on each iteration.

6.7 The Construct *skip*

The construct *skip* is true on an interval σ iff σ has length 1. We can express *skip* as follows:

$$skip \equiv_{\text{def}} \bigcirc empty.$$

Example (Measuring the length of an interval):
An interval's length can be tested using *skip* and *chop*. For example, the formula

$$skip; skip; skip$$

is true on intervals having length 3. It follows that this formula is equivalent to $len(3)$:

$$\models \quad len(3) \equiv (skip; skip; skip).$$

Example (Unit-length iterations):
The following while-loop simultaneously decrements I and sums I into J over each unit of time until I equals 0:

$$while\ (I \neq 0)\ do\ \bigl(skip \wedge [I \leftarrow I - 1] \wedge [J \leftarrow J + I]\bigr).$$

The body of the loop contains the *skip* operator in order that the length of each iterative step be 1. The behavior can also be expressed using *halt* and *gets*. Here is a semantically equivalent way of doing this:

$$halt(I = 0) \wedge (I\ gets\ I - 1) \wedge (J\ gets\ J + I).$$

Example (Expressing the operator *gets*):
A formula such as $A\ gets\ B$ can be alternatively expressed as an unspecified number of iterations, each of unit length:

$$\models \quad (A\ gets\ B) \equiv (skip \wedge [A \leftarrow B])^{*}.$$

6.8 Incorporating these Constructs into Tempura

We now extend Tempura to include the following statements based on the ITL constructs just introduced:

$skip$	(unit interval)
$w_1; w_2$	(sequential composition)
$for\ e\ times\ do\ w$	(simple for-loop)
$for\ v < e\ do\ w$	(indexed for-loop)
$while\ b\ do\ w$	(while-loop)
$repeat\ w\ until\ b$	(repeat-loop)
$loop\ w_1\ exit\ when\ b\ otherwise\ w_2$	(exit-loop)

Here w, w_1 and w_2 are themselves statements, e is an expression, v is a static variable and b is a boolean test.

Example (Computing sums):

The following Tempura program uses a while-loop to compute a sum:

$$(I = 4) \land (J = 0) \land \Box\ display(I, J) \\ \land \left[while\ I \neq 0\ do\ \left(skip \land [I \leftarrow I - 1] \land [J \leftarrow J + I] \right) \right]. \quad (6.1)$$

Figure 6.1 shows the program's behavior when run.

Example (Computing powers):

Consider the problem of finding the value of the expression I^J and placing it in another variable K. We can specify this using the temporal assignment

$$K \leftarrow I^J.$$

Figure 6.1: Execution of formula (6.1)

```
State  0:  I= 4   J= 0
State  1:  I= 3   J= 4
State  2:  I= 2   J= 7
State  3:  I= 1   J= 9
State  4:  I= 0   J=10

Done!  Computation length = 4.
```

The following Tempura statement achieves this by looking at J's binary structure:

$$(K = 1) \wedge \bigl[while\ (J > 0)\ do\ (skip \wedge w)\bigr],$$

where the statement w has the form

$$\begin{aligned}
&if\ (J\ mod\ 2 = 0)\\
&\quad then\ \bigl[(I \leftarrow I \cdot I) \wedge (J \leftarrow J \div 2) \wedge (K \leftarrow K)\bigr]\\
&\quad else\ \bigl[(I \leftarrow I) \wedge (J \leftarrow J - 1) \wedge (K \leftarrow K \cdot I)\bigr].
\end{aligned}$$

7 Some Applications

We now present some sample Tempura programs for summing the leaves of a tree, partitioning a list and sorting a list. Afterwards programs are given for simulating a simple multiplication circuit, generating digital pulses and testing a latch. The final two examples deal with synchronized communication between parallel processes. Most of the programming constructs used here have already been introduced. Those that have not are briefly described where mentioned.

7.1 Tree Summation

Suppose we have a binary tree of values such as either of the ones shown in figure 7.1. They can be linearly represented by the following list structures:

$$\langle\langle\langle 1,1\rangle,\langle 1,1\rangle\rangle,\langle\langle 1,1\rangle,\langle 1,1\rangle\rangle\rangle, \quad \langle\langle 1,\langle 2,3\rangle\rangle,\langle 4,5\rangle\rangle.$$

Let the function $leaf\_sum(tree)$ determine the sum of a tree's leaves:

$leaf\_sum(tree) \quad =_{\text{def}}$
$\quad if\ is\_integer(tree)\ then\ tree$
$\quad else\ leaf\_sum(tree_0) + leaf\_sum(tree_1).$

Here the predicate $is\_integer(tree)$ is true when the parameter $tree$ is an integer (i.e., a leaf) and false when $tree$ is a pair.

Now consider the task of designing an algorithm to reduce a tree in-place to a single value indicated by $leaf\_sum$. If the variable $Tree$ initially equals such a binary tree, we can specify the problem as follows:

$$Tree \leftarrow leaf\_sum(Tree).$$

Let us look at a serial and a parallel implementation of this.

7.1.1 Serial solution

One approach to achieving our goal is given by the predicate *serial_sum_tree* defined in figure 7.2. This terminates if the tree is already an integer-valued leaf. Otherwise, the predicate *sum_subtree* is used to first reduce the left subtree and then the right subtree. Finally, the statement

$$skip \land (Tree \leftarrow Tree_0 + Tree_1)$$

is used to reduce the tree to a single value. Note that when either of the two subtrees is being reduced by *sum_subtree*, the other one is kept stable. In addition, the built-in predicate *stable_struct* is used in the predicate *sum_subtree* to maintain the tree's main

Figure 7.1: Two binary trees

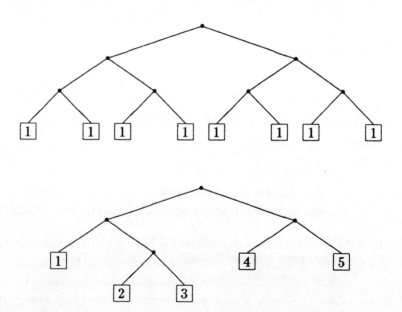

node intact so that the two subtrees can continue to be properly accessed. We can express *stable_struct* as follows:

$$stable\_struct\ e \quad \equiv_{\text{def}} \quad \Box\bigl[more \supset list([\bigcirc e], |e|)\bigr].$$

Thus, if the expression e is initially a list of some particular length, it remains a list of that length throughout the interval. If this operation were omitted, there would be no mention of whether the root of the tree remains a pair.

We initialize the tree and invoke *serial_sum_tree* by means of the following sort of formula:

$$Tree = \langle\langle\langle 1,1\rangle,\langle 1,1\rangle\rangle,\langle\langle 1,1\rangle,\langle 1,1\rangle\rangle\rangle$$
$$\land\ serial\_sum\_tree(Tree) \land \Box\ display(Tree).$$

Figure 7.3 shows the resulting behavior for the two different trees mentioned.

7.1.2 Parallel solution

The predicate *par_sum_tree* in figure 7.4 is similar to *serial_sum_tree* except that it recursively reduces each half of a pair in parallel rather than sequentially. A variable named *Done* is

Figure 7.2: Predicates to sequentially sum a binary tree

$$define\ serial\_sum\_tree(Tree) \equiv$$
$$\quad if\ is\_integer(Tree)\ then\ empty$$
$$\quad else\ ($$
$$\quad\quad sum\_subtree(Tree,0);$$
$$\quad\quad sum\_subtree(Tree,1);$$
$$\quad\quad ($$
$$\quad\quad\quad skip \land (Tree \leftarrow Tree_0 + Tree_1)$$
$$\quad\quad)$$
$$\quad).$$

$$define\ sum\_subtree(Tree,i) \equiv$$
$$\quad serial\_sum\_tree(Tree_i)$$
$$\quad \land\ stable\_struct\ Tree$$
$$\quad \land\ stable\ Tree_{1-i}.$$

Figure 7.3: Two executions of serial tree summation

```
State  0: Tree=<<<1,1>,<1,1>>,<<1,1>,<1,1>>>
State  1: Tree=<<2,<1,1>>,<<1,1>,<1,1>>>
State  2: Tree=<<2,2>,<<1,1>,<1,1>>>
State  3: Tree=<4,<<1,1>,<1,1>>>
State  4: Tree=<4,<2,<1,1>>>
State  5: Tree=<4,<2,2>>
State  6: Tree=<4,4>
State  7: Tree=8

Done!  Computation length = 7.


State  0: Tree=<<1,<2,3>>,<4,5>>
State  1: Tree=<<1,5>,<4,5>>
State  2: Tree=<6,<4,5>>
State  3: Tree=<6,9>
State  4: Tree=15

Done!  Computation length = 4.
```

Figure 7.4: Predicates to sum a binary tree in parallel

$define\ par\_sum\_tree(Tree) \equiv$
$\quad if\ is\_integer(Tree)\ then\ empty$
$\quad else\ ($
$\quad\quad \exists Done:($
$\quad\quad\quad (Done \approx [is\_integer(Tree_0) \wedge is\_integer(Tree_1)])$
$\quad\quad\quad \wedge\ halt\ Done$
$\quad\quad\quad \wedge\ stable\_struct\ Tree$
$\quad\quad\quad \wedge\ sum\_tree\_process(Done, Tree_0)$
$\quad\quad\quad \wedge\ sum\_tree\_process(Done, Tree_1)$
$\quad\quad);$
$\quad\quad ($
$\quad\quad\quad skip \wedge (Tree \leftarrow Tree_0 + Tree_1)$
$\quad\quad)$
$\quad).$

$define\ sum\_tree\_process(Done, Tree) \equiv$
$\quad process($
$\quad\quad par\_sum\_tree(Tree);$
$\quad\quad ($
$\quad\quad\quad halt\ Done$
$\quad\quad\quad \wedge\ stable\ Tree$
$\quad\quad)$
$\quad).$

used to monitor the progress of the two subtrees. It equals *true* when they are both finally integers. At this time the sum of the two values can be assigned to the tree variable. The subordinate predicate *sum_tree_process*(*Done, Tree*) reduces its tree parameter and then keeps the tree stable until the flag *Done* becomes true. This ensures that the two parallel invocations of *sum_tree_process* finish at the same time. Figure 7.5 shows the behavior of *par_sum_tree* on the two trees discussed above. As might be expected, the computation length is less than that required by the serial algorithm.

Note that the body of the predicate *sum_tree_process* in figure 7.4 is a statement of the form *process w*. The *process* construct has no special logical semantics:

$$process\ w \quad \equiv_{def} \quad w.$$

It is used when several Tempura statements are run in parallel and each independently determines the interval length. See section 9.3 for a discussion.

Figure 7.5: Two executions of parallel tree summation

```
State  0: Tree=<<<1,1>,<1,1>>,<<1,1>,<1,1>>>
State  1: Tree=<<2,2>,<2,2>>
State  2: Tree=<4,4>
State  3: Tree=8

Done!  Computation length = 3.


State  0: Tree=<<1,<2,3>>,<4,5>>
State  1: Tree=<<1,5>,9>
State  2: Tree=<6,9>
State  3: Tree=15

Done!  Computation length = 3.
```

7.1.3 Correctness and performance

ITL can be used to specify the correctness and relative speeds of the two algorithms just introduced. Here are the basic correctness properties:

$$\models \quad serial\_sum\_tree(Tree) \supset \bigl(Tree \leftarrow leaf\_sum(Tree)\bigr),$$
$$\models \quad par\_sum\_tree(Tree) \supset \bigl(Tree \leftarrow leaf\_sum(Tree)\bigr).$$

The invariant and rate of progress for the serial version are shown below:

$$\models \quad serial\_sum\_tree(Tree) \supset$$
$$\bigl(stable\ leaf\_sum(Tree)$$
$$\land\ [leaf\_count(Tree)\ gets\ leaf\_count(Tree) - 1]\bigr).$$

Here the function *leaf_count* equals the number of leaves in a tree. The property states that even though the tree is changing, the sum of its leaves remains stable. Furthermore, the number of leaves decreases by 1 every unit of time. The invariant for the parallel algorithm is identical to that of the serial algorithm although the rate of progress is more complicated. The tree's leaf count decreases by the number of internal nodes whose two sons are both integers.

The computation length for the serial and parallel algorithms is expressed as follows:

$$\models \quad serial\_sum\_tree(Tree) \supset len(leaf\_count(Tree) - 1),$$
$$\models \quad par\_sum\_tree(Tree) \supset len(tree\_height(Tree)).$$

Here the expression $leaf\_count(Tree) - 1$ equals the number of internal nodes in the tree. The function *tree_height* is defined to equal the length of the longest path from the tree's root to a leaf:

$$tree\_height(tree) \quad =_{\text{def}}$$
$$if\ is\_integer(tree)\ then\ 0$$
$$else\ 1 + max\bigl(tree\_height(tree_0), tree\_height(tree_1)\bigr).$$

7.2 Partitioning a List

We now describe a technique for partitioning a list. This will be subsequently used in some quicksort algorithms. The predicate

$$partition\_list(L, key, left\_len)$$

defined in figure 7.6 uses the predicate *part_loop* to iteratively permute the elements of the list parameter L so that those less than the value of *key* end up to the left of elements greater than or equal to *key*. Each step of the loop invokes the predicate *part_step*. The length of the left side is finally stored in the static parameter *left_len*. For example, suppose that L initially equals the list $\langle 1, 3, 2, 3, 0, 1, 3 \rangle$ and *key* has the value 2. The final value of L is then the list $\langle 1, 1, 0, 3, 2, 3, 3 \rangle$ and the value of *left_len* is 3. In figure 7.7 we depict the behavior of L in each state and display the value of *left_len* in the final state. Note that the predicate *part_step* references a predicate called *swap_list*. This has the general form

$$swap\_list(L, i, j)$$

and exchanges the values of L_i and L_j, leaving the remaining elements of the list L unchanged. Here is one way to express this in ITL:

$$\forall k < |L|:\\ \big[\textit{if } k = i \textit{ then } (L_k \leftarrow L_j) \\ \textit{else if } k = j \textit{ then } (L_k \leftarrow L_i) \textit{ else } (L_k \leftarrow L_k) \big].$$

7.2.1 Correctness of *partition_list*

The correctness of this algorithm can be expressed using two properties. The first one states that the list variable L's final value is a permutation of its initial one:

$$\models \quad partition\_list(L, key, left\_len) \\ \supset [list\_to\_bag(L) \leftarrow list\_to\_bag(L)].$$

Here the function $list\_to\_bag(L)$ equals the bag (multi-set) corresponding to L. Thus we express the fact that the initial and final bags for L are identical.

Figure 7.6: Predicates for partitioning a list

$\text{define } partition\_list(L, key, left\_len) \equiv$
$\quad \exists I, J : ($
$\quad\quad (I = 0) \wedge (J = |L|)$
$\quad\quad \wedge \, ($
$\quad\quad\quad part\_loop(L, key, I, J);$
$\quad\quad\quad [empty \wedge (left\_len = I)]$
$\quad\quad)$
$\quad).$

$\text{define } part\_loop(L, key, I, J) \equiv$
$\quad \text{while } I < J \text{ do } ($
$\quad\quad skip \wedge part\_step(L, key, I, J)$
$\quad).$

$\text{define } part\_step(L, key, I, J) \equiv$
$\quad \text{if } L_I < key$
$\quad \text{then } ($
$\quad\quad stable \ L$
$\quad\quad \wedge (I \leftarrow I + 1) \wedge (J \leftarrow J)$
$\quad)$
$\quad \text{else } ($
$\quad\quad swap\_list(L, I, J - 1)$
$\quad\quad \wedge (I \leftarrow I) \wedge (J \leftarrow J - 1)$
$\quad).$

The second property states that L's final value is partitioned according to *key*:

$$\models \quad \textit{partition\_list}(L, \textit{key}, \textit{left\_len}) \\ \supset \textit{fin}\big(\forall i < |L| : \big[(i < \textit{left\_len}) \equiv (L_i < \textit{key})\big]\big).$$

The definition of *partition_list* uses the predicate *part_loop* which iterates over L while the variables I and J index the start and end of the sublist of elements not yet processed. The following property states that *part_loop* leaves the multi-set representation of L stable:

$$\models \quad \textit{part\_loop}(L, \textit{key}, I, J) \supset \textit{stable list\_to\_bag}(L).$$

In addition, throughout the computation the elements L_0, ..., L_{I-1} are all less than the key and the elements L_J, L_{J+1}, ..., $L_{|L|-1}$ are greater than or equal to the key:

$$\models \quad \textit{part\_loop}(L, \textit{key}, I, J) \supset \\ \big[\Box \forall k < I \colon (L_k < \textit{key}) \\ \wedge \Box \forall k \colon \big([J \leq k < |L|] \supset [L_k \geq \textit{key}]\big)\big].$$

Eventual termination of *part_loop* is indicated by the fact that in every step the value of J is greater than or equal to that

Figure 7.7: Execution of predicate for partitioning

```
State  0:  L=<1,3,2,3,0,1,3>
State  1:  L=<1,3,2,3,0,1,3>
State  2:  L=<1,3,2,3,0,1,3>
State  3:  L=<1,1,2,3,0,3,3>
State  4:  L=<1,1,2,3,0,3,3>
State  5:  L=<1,1,0,3,2,3,3>
State  6:  L=<1,1,0,3,2,3,3>
State  7:  L=<1,1,0,3,2,3,3>
State  7:  left_len= 3

Done!  Computation length = 7.
```

of I and either I increases or J decreases by 1. Therefore, the difference $J - I$ continues to decrease by 1:

$$\models \quad part\_loop(L, key, I, J) \supset [(J - I) \text{ gets } (J - I - 1)].$$

Since I is initially 0 and J is initially $|L|$, it follows that $|L|$ units of time are required for the entire computation:

$$\models \quad part\_loop(L, key, I, J) \supset len(|L|).$$

7.3 Quicksort

Let $sort(e)$ be a function equalling the list expression e's sorted value. We can then specify the in-place sort of a list variable L by means of the formula

$$L \leftarrow sort(L).$$

Another way to express this is as follows:

$$[list\_to\_bag(L) \leftarrow list\_to\_bag(L)] \wedge [fin\ sorted(L)].$$

Here the predicate $sorted(e)$ is true if the list expression e is sorted. Thus, the overall formula states that L's final value is a permutation of its original one since the bag corresponding to L remains unchanged. Furthermore, L's final value is in sorted order. One way to sort a list is by using the predicate *serial_quicksort* shown in figure 7.8. We use a special subscripting construct of the form $e_1[e_2 \mathinner{.\,.} e_3]$. This is a sublist of the expression e_1 of length $e_3 - e_2$ and has the form

$$\langle e_1[e_2], e_1[e_2 + 1], \ldots, e_1[e_3 - 1]\rangle.$$

It can be expressed as follows

$$e_1[e_2 \mathinner{.\,.} e_3] \quad =_{\text{def}} \quad \langle e_1[i]: e_2 \leq i < e_3\rangle,$$

where the static variable i does not occur freely in e_1. Note that $e_1[e_3]$ is not included among the sublist's elements. We generally write a sublist expression such as $L[0 \mathinner{.\,.} k]$ in the form $L_{0..k}$.

The quicksort algorithm presented here leaves the list unchanged if it has 0 or 1 elements and otherwise partitions it into

Figure 7.8: Predicates for serial quicksort

$define\ serial\_quicksort(L) \equiv$
$if\ |L| \leq 1\ then\ empty$
$else\ \exists pivot:($
   $quick\_partition(L, pivot);$
   $serial\_sort\_parts(L, pivot)$
$).$

$define\ serial\_sort\_parts(L, k) \equiv$
$($
   $serial\_quicksort(L_{0..k})$
   $\wedge\ stable\ L_{k..|L|}$
$);$
$($
   $serial\_quicksort(L_{(k+1)..|L|})$
   $\wedge\ stable\ L_{0..(k+1)}$
$).$

$define\ quick\_partition(L, pivot) \equiv$
$($
   $partition\_list(L_{0..|L|-1}, L_{|L|-1}, pivot)$
   $\wedge\ stable\ L_{|L|-1}$
$);$
$($
   $skip \wedge swap\_list(L, pivot, |L|-1)$
$).$

two main parts with a pivot element in between. The left half has elements less than the pivot and the right half has elements greater than or equal to the pivot. The left part is then recursively sorted, after which the right part is itself sorted. Throughout this time, the pivot is kept stable. The partitioning operation is performed by the predicate *quick_partition* and the subsequent sorting of the two parts is achieved by the predicate *serial_sort_parts*. Both of these are described below.

Here is a program for testing the quicksort predicate:

$$fixed\_list(L, 7) \wedge fixed\_list(T, 7) \wedge \Box \, display(L, T)$$
$$\wedge \, (L = \langle 4, 5, 2, 0, 6, 1, 3 \rangle) \wedge serial\_quicksort(L)$$
$$\wedge \, \forall i < |L| : [T_i \approx (if \; i = L_i \; then \; 1 \; else \; 0)].$$

The statement $fixed\_list(L, 7)$ specifies that L is always a list of length 7. It is logically equivalent to the formula

$$\Box \, list(L, 7)$$

but turns out to be more natural and much more efficient to use in Tempura. We include a list variable T that has the same length as L and shows which elements of L are in their proper positions. For any $i < |L|$, the i-th element of T has the value 1 in a state iff the value of L_i in that state equals i. Of course, this technique only works if L's initial value is a permutation of the integers 0, 1, ..., $|L| - 1$. Figure 7.9 shows the program's execution. The list L itself is partitioned from state 0 to state 7 with the value 3 used as the key. In particular, from state 6 to state 7, the value 3 is moved to its proper position in L_3. From state 7 to state 12, the left sublist $L_{0..3}$ (i.e., L_0, L_1 and L_2) is sorted. From state 12 to state 17, the right sublist $L_{4..7}$ (i.e., L_4, L_5 and L_6) is sorted.

7.3.1 Explanation of *quick_partition*

The predicate $quick\_partition(L, pivot)$ permutes L and assigns the static parameter *pivot* an index into L so that the elements L_0, ..., $L_{pivot-1}$ are all less than L_{pivot} and the elements $L_{pivot+1}$, ..., $L_{|L|-1}$ are all greater than or equal to L_{pivot}. This is achieved by first invoking the predicate *partition_list* described previously in section 7.2 on the sublist $L_0, ..., L_{|L|-2}$ with the

Figure 7.9: Execution of serial quicksort

```
State  0: L=<4,5,2,0,6,1,3>   T=<0,0,1,0,0,0,0>
State  1: L=<1,5,2,0,6,4,3>   T=<0,0,1,0,0,0,0>
State  2: L=<1,5,2,0,6,4,3>   T=<0,0,1,0,0,0,0>
State  3: L=<1,6,2,0,5,4,3>   T=<0,0,1,0,0,0,0>
State  4: L=<1,0,2,6,5,4,3>   T=<0,0,1,0,0,0,0>
State  5: L=<1,0,2,6,5,4,3>   T=<0,0,1,0,0,0,0>
State  6: L=<1,0,2,6,5,4,3>   T=<0,0,1,0,0,0,0>
State  7: L=<1,0,2,3,5,4,6>   T=<0,0,1,1,0,0,1>
State  8: L=<1,0,2,3,5,4,6>   T=<0,0,1,1,0,0,1>
State  9: L=<1,0,2,3,5,4,6>   T=<0,0,1,1,0,0,1>
State 10: L=<1,0,2,3,5,4,6>   T=<0,0,1,1,0,0,1>
State 11: L=<1,0,2,3,5,4,6>   T=<0,0,1,1,0,0,1>
State 12: L=<0,1,2,3,5,4,6>   T=<1,1,1,1,0,0,1>
State 13: L=<0,1,2,3,5,4,6>   T=<1,1,1,1,0,0,1>
State 14: L=<0,1,2,3,5,4,6>   T=<1,1,1,1,0,0,1>
State 15: L=<0,1,2,3,5,4,6>   T=<1,1,1,1,0,0,1>
State 16: L=<0,1,2,3,5,4,6>   T=<1,1,1,1,0,0,1>
State 17: L=<0,1,2,3,4,5,6>   T=<1,1,1,1,1,1,1>

Done!  Computation length = 17.
```

rightmost element $L_{|L|-1}$ acting as key. The value of $L_{|L|-1}$ is itself keep stable. Afterwards, the value of *pivot* is an index to the start of the second half of the partition. The element $L_{|L|-1}$ is then exchanged with L_{pivot}.

7.3.2 Explanation of *serial_sort_parts*

The predicate *serial_sort_parts*(L, *pivot*) first sorts the left partition $L_0, \ldots, L_{pivot-1}$ and after this the right partition $L_{pivot+1}, \ldots, L_{|L|-1}$. During the sorting of either side, the other remains stable. Throughout the entire period the value of the pivot element L_{pivot} is left unchanged since it is already in its proper position.

7.3.3 Parallel quicksort

The predicate *par_quicksort* shown in figure 7.10 is a parallel variant of the serial algorithm just described. As the execution in figure 7.11 demonstrates, this version of quicksort can take fewer computational steps than the serial one. The basic change to the sorting technique is seen in the predicate *par_sort_parts*. This sorts the two sublists of L in parallel rather than in succession.

Each sublist has a flag (i.e., *Ready1* or *Ready2*) associated with an invocation of the predicate *sort_process*. This predicate recursively sorts the sublist and then sets the flag to true indicating completion. It then waits for the flag *Done*, which becomes true when both sublists are finished. In the meantime, the sublist is kept stable.

7.4 A Multiplication Circuit

Figure 7.12 depicts a simple multiplication circuit containing a number of component types such as *flipflop*, *zero_test* and *adder*. This was originally designed and verified by Mike Gordon using the LSM behavioral notation [12]. In figure 7.13 we define the components as predicates in ITL. Gordon used the boolean values *true* and *false* to represent bit signals. We follow the same convention as can be seen in such devices as *zero_test* and *or_gate*. The components *reg* and *flipflop* in the original specification both provide a form of unit delay and can be modelled by the ITL construct *gets*. The overall multiplier is then given in figure 7.14.

Figure 7.10: Predicates for parallel quicksort

$define\ par\_quicksort(L) \equiv$
$if\ |L| \leq 1\ then\ empty$
$else\ \exists pivot:($
   $quick\_partition(L, pivot);$
   $par\_sort\_parts(L, pivot)$
$).$

$define\ par\_sort\_parts(L, pivot) \equiv$
$\exists Done, Ready1, Ready2:($
   $(Done \approx (Ready1 \wedge Ready2))$
   $\wedge\ halt\ Done$
   $\wedge\ sort\_process(Done, Ready1, L_{0..pivot})$
   $\wedge\ sort\_process(Done, Ready2, L_{pivot+1..|L|})$
$)$
$\wedge\ stable\ L_{pivot}.$

$define\ sort\_process(Done, Ready, L) \equiv$
$process($
   $($
     $par\_quicksort(L) \wedge (Ready \approx empty)$
   $);$
   $($
     $(halt\ Done) \wedge (stable\ L) \wedge (stable\ Ready)$
   $)$
$).$

Various internal signals such as *B1* and *L3* are hidden by means of existential quantification ($\exists$). The signals *in1*, *in2*, *Done* and *Out* are left accessible. The device's structure is represented as a conjunction of instances of the individual components. In addition, equalities are included to properly initialize the signals *Done*, *Out* and *L2*.

In order to test the multiplier, we feed the circuit some data and then terminate when the circuit has the answer. The following program performs these tasks and has the multiplier determine the product of 4 and 9:

$mult\_imp(4, 9, Done, Out)$
$\land (\bigcirc halt\ Done) \land \square\ display(Done, Out).$

The execution of this is shown in figure 7.15. Note that the flag *Done* is initialized to *true* and the simulation halts the next time it becomes true.

7.5 Pulse Generation

We now demonstrate how Tempura can be used to generate, manipulate and display simple digital waveforms. Let us first

Figure 7.11: Execution of parallel quicksort

```
State  0: L=<4,5,2,0,6,1,3>   T=<0,0,1,0,0,0,0>
State  1: L=<1,5,2,0,6,4,3>   T=<0,0,1,0,0,0,0>
State  2: L=<1,5,2,0,6,4,3>   T=<0,0,1,0,0,0,0>
State  3: L=<1,6,2,0,5,4,3>   T=<0,0,1,0,0,0,0>
State  4: L=<1,0,2,6,5,4,3>   T=<0,0,1,0,0,0,0>
State  5: L=<1,0,2,6,5,4,3>   T=<0,0,1,0,0,0,0>
State  6: L=<1,0,2,6,5,4,3>   T=<0,0,1,0,0,0,0>
State  7: L=<1,0,2,3,5,4,6>   T=<0,0,1,1,0,0,1>
State  8: L=<1,0,2,3,5,4,6>   T=<0,0,1,1,0,0,1>
State  9: L=<1,0,2,3,5,4,6>   T=<0,0,1,1,0,0,1>
State 10: L=<1,0,2,3,5,4,6>   T=<0,0,1,1,0,0,1>
State 11: L=<1,0,2,3,5,4,6>   T=<0,0,1,1,0,0,1>
State 12: L=<0,1,2,3,4,5,6>   T=<1,1,1,1,1,1,1>

Done! Computation length = 12.
```

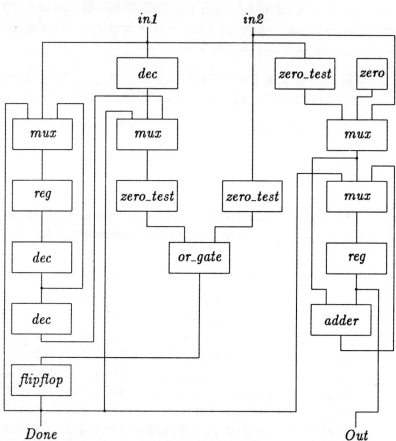

Figure 7.12: Block diagram of multiplication circuit

extend the boolean operators ¬, ∧ and ∨ to permit the bit values
0 and 1. For example the value of the expression ¬0 is 1 and the
value of $1 \wedge 0$ is 0. We use these constructs in the program in figure 7.16. This controls the behavior of the four bit variables W,
X, Y and Z. The signal W is initialized to 0 and then successively
oscillates 5 times over 4-unit intervals by means of the following
form of sequential iteration:

> *for* 5 *times do* (
> $[len(3) \wedge stable\ W]; [skip \wedge (W \leftarrow \neg W)]$
>).

The total length of the period is therefore 20 units. In parallel
with this, the signal X is initialized to 0 and then receives the
values of W but with unit delay. The same happens from X to
Y. The value of Z is always the bit-and of W, X and Y. In
figure 7.17, we display the behavior of the combined system in
the form of a timing diagram. This is best viewed when turned

Figure 7.13: Definitions of individual components

> *define* $mux(Switch, In1, In2, Out) \equiv$
> $\Box \big(if\ Switch\ then\ Out = In1\ else\ Out = In2 \big)$.

> *define* $reg(In, Out) \equiv (Out\ gets\ In)$.

> *define* $flipflop(In, Out) \equiv (Out\ gets\ In)$.

> *define* $dec(In, Out) \equiv$
> $Out \approx (if\ In = 0\ then\ 0\ else\ In - 1)$.

> *define* $adder(In1, In2, Out) \equiv (Out \approx [In1 + In2])$.

> *define* $zero\_test(In, Out) \equiv (Out \approx [In = 0])$.

> *define* $or\_gate(In1, In2, Out) \equiv (Out \approx [In1 \vee In2])$.

> *define* $zero(Out) \equiv (Out \approx 0)$.

Figure 7.14: Definition of multiplier

$define\ mult\_imp(in1, in2, Done, Out) \equiv$
$\quad \exists B1, B2, B3, B4, L1, L2, L3, L4, L5, L6, L7, L8, L9, L10 : ($
$\quad\quad mux(Done, L9, L8, L7)$
$\quad\quad \wedge reg(L7, Out)$
$\quad\quad \wedge adder(L9, Out, L8)$
$\quad\quad \wedge dec(in1, L6)$
$\quad\quad \wedge mux(Done, L6, L4, L5)$
$\quad\quad \wedge mux(Done, in1, L3, L1)$
$\quad\quad \wedge reg(L1, L2)$
$\quad\quad \wedge dec(L2, L3)$
$\quad\quad \wedge dec(L3, L4)$
$\quad\quad \wedge zero(L10)$
$\quad\quad \wedge mux(B4, L10, in2, L9)$
$\quad\quad \wedge zero\_test(in1, B4)$
$\quad\quad \wedge zero\_test(L5, B1)$
$\quad\quad \wedge zero\_test(in2, B2)$
$\quad\quad \wedge or\_gate(B1, B2, B3)$
$\quad\quad \wedge flipflop(B3, Done)$
$\quad\quad \wedge (Done = true) \wedge (Out = 0) \wedge (L2 = 0)$
$).$

Figure 7.15: Simulation of multiplier

```
State 0: Done=true   Out= 0
State 1: Done=false  Out= 9
State 2: Done=false  Out=18
State 3: Done=false  Out=27
State 4: Done=true   Out=36

Done! Computation length = 4.
```

sideways. The style of output used here is not hard to generate in Tempura although we omit the details.

7.6 Testing a Latch

The device shown in figure 7.18 is a simple latch built out of two cross-coupled nor-gates. When the bit inputs S and R are held stable long enough, the outputs Q and $\overline{Q}$ respond to them according to the table in figure 7.19. If S and R are both 0, the device retains its current state. The values of the inputs should not be held simultaneously at 1 since this can result in the latch's outputs later oscillating.

The program in figure 7.20 simulates the latch for values of the inputs S and R. We model each nor-gate as having unit delay. Note that the variable $\overline{Q}$ is referred to as *Qbar* in the program. The resulting system behavior is displayed in figure 7.21. The iterative construct

$$\textit{for } v \in e \textit{ do } w$$

is used to sequentially assign a locally scoped variable v the elements of the given list expression e and execute the statement w with each such binding.

7.7 Synchronized Communication

Tempura can be used to model parallel processes that periodically send or receive data from one another. Figure 7.22 shows a block diagram containing two modules with a connection between them. The left module transmits information to the right one by means of a *stream* package that we have implemented in

Figure 7.16: Program to generate some digital waveforms

$$(W = 0) \wedge (X = 0) \wedge (Y = 0) \wedge$$
$$\textit{for } 5 \textit{ times do } ($$
$$\quad [len(3) \wedge (\textit{stable } W)];$$
$$\quad [\textit{skip} \wedge (W \leftarrow \neg W)]$$
$$)$$
$$\wedge (X \textit{ gets } W) \wedge (Y \textit{ gets } X)$$
$$\wedge (Z \approx [W \wedge X \wedge Y]).$$

Figure 7.17: Execution of waveform generator

```
State  0:  Z    Y    X    W
State  0:  |    |    |    |
State  1:  |    |    |    |
State  2:  |    |    |    |
State  3:  |    |    |    |
State  4:  |    |    |         |
State  5:  |    |         |    |
State  6:  |         |    |    |
State  7:  |         |    |    |
State  8:  |         |    |    |
State  9:  |         |    |    |
State 10:  |         |    |    |
State 11:  |         |    |    |
State 12:  |         |    |         |
State 13:  |         |         |    |
State 14:  |    |         |    |
State 15:  |    |         |    |
State 16:  |    |         |    |
State 17:  |         |    |    |
State 18:  |    |    |    |
State 19:  |    |    |    |
State 20:  |    |    |         |
```

Done! Computation length = 20.

Figure 7.18: Block diagram of SR-latch

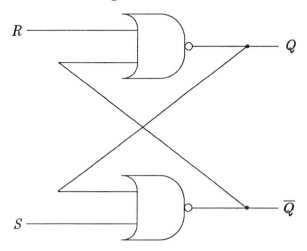

Figure 7.19: SR-latch operation

operation	S	R	Q	$\overline{Q}$
set to 1	1	0	1	0
clear to 0	0	1	0	1
no change	0	0	old Q	old $\overline{Q}$
undefined	1	1	–	–

Figure 7.20: Program to simulate SR-latch

$(S = 0) \wedge (R = 0) \wedge (Q = 0) \wedge (Qbar = 0) \wedge$
$for\ l \in \langle \langle 1,0 \rangle, \langle 0,0 \rangle, \langle 0,1 \rangle, \langle 1,0 \rangle, \langle 0,0 \rangle \rangle$
$do\ ($
$\quad len(5) \wedge (S\ gets\ l_0) \wedge (R\ gets\ l_1)$
$)$
$\wedge (Q\ gets\ \neg[R \vee Qbar])$
$\wedge (Qbar\ gets\ \neg[S \vee Q]).$

Figure 7.21: Simulation of SR-latch

```
State  0: Qbar   Q     R     S
State  0:  |     |     |     |
State  1:  |     |     |     |
State  2:  |     |     |     |
State  3:  |     |     |     |
State  4:  |     |     |     |
State  5:  |     |     |     |
State  6:  |     |     |     |
State  7:  |     |     |     |
State  8:  |     |     |     |
State  9:  |     |     |     |
State 10:  |     |     |     |
State 11:  |     |     |     |
State 12:  |     |     |     |
State 13:  |     |     |     |
State 14:  |     |     |     |
State 15:  |     |     |     |
State 16:  |     |     |     |
State 17:  |     |     |     |
State 18:  |     |     |     |
State 19:  |     |     |     |
State 20:  |     |     |     |
State 21:  |     |     |     |
State 22:  |     |     |     |
State 23:  |     |     |     |
State 24:  |     |     |     |
State 25:  |     |     |     |

Done!  Computation length = 25.
```

Figure 7.22: Block diagram of sender-receiver system

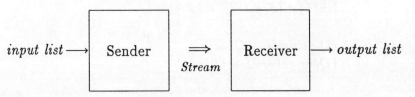

Tempura. We omit the implementation details but give a sample execution of the system in figure 7.23. In states 3, 6, 9, 12 and 15 a value is passed on. The end-of-stream marker is passed in state 18. In the final state, the receiver shows that it has successfully accepted the five data messages.

In figure 7.24 is a block diagram of a system that does parallel lexical analysis, parsing and evaluation of a string of characters representing an arithmetic expression. The heart of the corresponding Tempura program is shown in figure 7.25. Here four processes are connected together by means of streams. The first one takes the string and feeds it character-by-character into the lexical analyzer. This simultaneously outputs tokens to the parser. This in turn outputs reverse Polish notation to the evaluator. Eventually the evaluator determines the expression's numeric value. An execution of the system processing the string "10␣+2␣" is shown in figure 7.26. We use the character ␣ to visibly represent a space. Note that in states 13 and 24 two communications occur at once.

Figure 7.23: Execution of sender-receiver system

```
State  0: Stream_status
State  0: <>
State  1: <>
State  2: <>
State  3: <<0>>
State  4: <>
State  5: <>
State  6: <<10>>
State  7: <>
State  8: <>
State  9: <<20>>
State 10: <>
State 11: <>
State 12: <<30>>
State 13: <>
State 14: <>
State 15: <<40>>
State 16: <>
State 17: <>
State 18: <<>>
State 19: <>
State 20: <>
State 21: <>
State 22: <>
State 23: <>
State 24: <>
State 25: <>
State 26: <>
State 26: output list = <0,10,20,30,40>

Done!   Computation length = 26.
```

Figure 7.24: Block diagram of parallel lexer-parser-evaluator

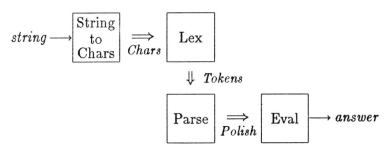

Figure 7.25: Heart of parallel lexer-parser-evaluator

$\ldots$
$\wedge$ *list_to_port_process*(Sig_list_0,
    string_input, *sender_port*(*Char_stream*))
$\wedge$ *lexer_process*(Sig_list_1,
    receiver_port(*Char_stream*),
    sender_port(*Token_stream*))
$\wedge$ *parser_process*(Sig_list_2,
    receiver_port(*Token_stream*),
    sender_port(*Polish_stream*))
$\wedge$ *eval_process*(Sig_list_3,
    receiver_port(*Polish_stream*), *answer*)
$\ldots$

Figure 7.26: Execution of lexer-parser-evaluator on "10␣+2␣"

```
State  0: Char         Token        Polish
State  0: <>           <>           <>
State  1: <>           <>           <>
State  2: <>           <>           <>
State  3: <<"1">>      <>           <>
State  4: <>           <>           <>
State  5: <>           <>           <>
State  6: <<"0">>      <>           <>
State  7: <>           <>           <>
State  8: <>           <>           <>
State  9: <<"␣">>      <>           <>
State 10: <>           <>           <>
State 11: <>           <<10>>       <>
State 12: <>           <>           <>
State 13: <<"+">>      <>           <<10>>
State 14: <>           <>           <>
State 15: <>           <<"+">>      <>
State 16: <>           <>           <>
State 17: <<"2">>      <>           <>
State 18: <>           <>           <>
State 19: <>           <>           <>
State 20: <<"␣">>      <>           <>
State 21: <>           <>           <>
State 22: <>           <<2>>        <>
State 23: <>           <>           <>
State 24: <<>>         <>           <<2>>
State 25: <>           <>           <>
State 26: <>           <<>>         <>
State 27: <>           <>           <>
State 28: <>           <>           <<"+">>
State 29: <>           <>           <>
State 30: <>           <>           <>
State 31: <>           <>           <<>>
State 32: <>           <>           <>
State 33: <>           <>           <>
State 34: <>           <>           <>
State 34: Final answer = 12
Done!  Computation length = 34.
```

8 An Interpreter for Tempura

Let us now consider how to build an interpreter for Tempura. The aim of the presentation is to give an idea of how to execute Tempura programs. Therefore a number of low-level aspects of the implementation are not described. This material can be skipped.

Before discussing the design of the system, we look at how the approach is applied to the following program:

$$(\bigcirc \bigcirc \; empty) \land (I = 0) \land (I \; gets \; I + 1) \land \Box (J = 2 \cdot I). \quad (8.1)$$

This is true on intervals of length 2 in which I assumes the successive values 0, 1, and 2 while J simultaneously assumes the values 0, 2, and 4.

One way to execute such a formula is to transform it to a logically equivalent conjunction of the two formulas *present_state* and $\circledcirc$ *what_remains*:

present_state $\land \circledcirc$ *what_remains*.

Here, the formula *present_state* consists of assignments to the program variables and also indicates whether or not the interval is finished. The formula *what_remains* is what is executed in subsequent states if the interval does indeed continue on. Thus, it can be viewed as a kind of continuation.

For the formula under consideration, *present_state* has the following value:

$(I = 0) \land (J = 0) \land$ *more*.

The value of *what_remains* is the formula

$$(\bigcirc empty) \land (I = 1) \land (I \ gets \ I + 1) \land \Box(J = 2 \cdot I).$$

In figure 8.1 we show the effects of such transformations before and after each of the three states of the execution.

8.1 Variables Used by the Interpreter

The operation of the Tempura interpreter is based on the technique just described. When executing a Tempura program, the interpreter makes use of the four variables *Program*, *Memory*, *Current_Env* and *Current_Done_Cell*. They are accessed by the interpreter's various routines. Below is a summary of each of these variables:

- *Program*: This is a variable that initially contains the Tempura program itself. After the execution of each state, it is transformed to a formula of the form $\circledcirc w$, where w describes what should be done in the next state.

- *Memory*: This is an indexed list of cells. Each cell can be empty or contain a value such as an integer or a list descriptor. At the beginning of each state, every cell is assigned the empty list $\langle\rangle$, thus indicating that no value is being stored. When a value c is to be placed in the cell, the cell's actual contents are assigned the singleton $\langle c \rangle$.

- *Current_Env*: This contains an environment, which is a list having a separate entry for each variable in the Tempura program. Each entry is a pair with the name of the associated variable as well as an index to a memory cell that holds the variable's value. For the Tempura program described earlier, the value of *Current_Env* might be the following:

$$\langle \langle "I", 0 \rangle, \langle "J", 1 \rangle \rangle.$$

- *Current_Done_Cell*: This equals the index of a memory cell called the *done-flag*. The Tempura program places either *true* or *false* in the done-flag during every state, thus

Figure 8.1: Transformation of formula (8.1)

Before state 0:
$(\bigcirc \bigcirc empty) \wedge (I = 0) \wedge (I \ gets \ I + 1) \wedge \square(J = 2 \cdot I)$
After state 0:
$[(I = 0) \wedge (J = 0) \wedge more]$
$\wedge \circledcirc [(\bigcirc empty) \wedge (I = 1) \wedge (I \ gets \ I + 1) \wedge \square(J = 2 \cdot I)]$

Before state 1:
$(\bigcirc empty) \wedge (I = 1) \wedge (I \ gets \ I + 1) \wedge \square(J = 2 \cdot I)$
After state 1:
$[(I = 1) \wedge (J = 2) \wedge more]$
$\wedge \circledcirc [empty \wedge (I = 2) \wedge (I \ gets \ I + 1) \wedge \square(J = 2 \cdot I)]$

Before state 2:
$empty \wedge (I = 2) \wedge (I \ gets \ I + 1) \wedge \square(J = 2 \cdot I)$
After state 2:
$[(I = 2) \wedge (J = 4) \wedge empty]$
$\wedge \circledcirc [false \wedge (I \ gets \ I + 1) \wedge \square(J = 2 \cdot I)]$

indicating whether or not that state is the final one. For
example, in a state where the statement *empty* is encountered, the interpreter puts the value *true* in the done-flag
cell. If the statement *more* is encountered, the value *false*
is used instead. If the user fails to assign a value to the
done-flag in a particular state, the interpreter detects this
and flags an error.

8.2 Basic Execution Algorithm

The basic algorithm used by the interpreter can be represented in the following procedural form:

begin
   local Program, Memory, Current_Env, Current_Done_Cell;
   prepare_execution_of_program;
   loop
     execute_single_state
     exit when Memory[*Current_Done_Cell*] = $\langle true \rangle$
     otherwise
       advance_to_next_state
end.

Here is a more detailed look at each part of the process:

- *prepare_execution_of_program*: The interpreter's variable *Program* is assigned the program's syntax tree and the variable *Current_Env* is initialized to indicate suitable references into the memory. The memory itself is allocated to have one cell for each program variable as well as a cell whose index is placed in *Current_Done_Cell*. All the memory cells are initially emptied (i.e., set to $\langle \rangle$).

- *execute_single_state*: The value of the variable *Program* is transformed until it is of the form $\odot\, w$. A check is made to ensure that the done-flag indexed by *Current_Done_Cell* has been set to *true* or *false*. The actual transformations used for each Tempura construct are described later on. All assignments occurring in the current state are reflected in the values of the memory's cells.

- *advance_to_next_state*: If the current state is not the last, preparations are made for processing the next one. This is done by emptying the contents of all the memory cells and deleting the leading operator Ⓦ from the formula held in *Program*.

Figure 8.2 shows the details of executing the simple program (8.1).

8.3 Description of the Procedure *execute_single_state*

We now define the procedure *execute_single_state*:

procedure execute_single_state:
begin
   while ¬*is_reduced_stmt*(*Program*) *do*
     transform_stmt(*Program*);
   if Memory[*Current_Done_Cell*] = $\langle\rangle$ *then*
     error(*"Termination status not specified."*)
end.

The while-loop repeatedly transforms the program using the procedure *transform_stmt* until the test

   is_reduced_stmt(*Program*)

is true, signifying that the program is reduced to the form Ⓦ w. Each iteration of the loop corresponds to one pass over the program. Sometimes many passes can be required to completely process one state. Once this is achieved, no further transformations take place in that state. A check is then made to ensure that the done-flag has been properly set. The operation of *transform_stmt* is explained in detail below.

8.4 Description of the Procedure *transform_stmt*

The procedure *transform_stmt* has the form

   transform_stmt(*Statement*).

It makes a single left-to-right pass over a Tempura statement held in the parameter *Statement* and transforms it, while simultaneously extracting information about Tempura variables and termination. The contents of the memory cells are updated in the

Figure 8.2: Details of execution of formula (8.1)

Current_Env: $\langle\langle"I",0\rangle,\langle"J",1\rangle\rangle$,
Current_Done_Cell: 2.

Before state 0:
 Program:
  $(\bigcirc\bigcirc empty) \wedge (I=0) \wedge (I\ gets\ I+1) \wedge \square(J=2\cdot I)$,
 Memory: $\langle\langle\rangle,\langle\rangle,\langle\rangle\rangle$.
After state 0:
 Program:
  $\circledcirc\big[(\bigcirc empty) \wedge (I=1) \wedge (I\ gets\ I+1) \wedge \square(J=2\cdot I)\big]$,
 Memory: $\langle\langle 0\rangle,\langle 0\rangle,\langle false\rangle\rangle$.

Before state 1:
 Program: $(\bigcirc empty) \wedge (I=1) \wedge (I\ gets\ I+1) \wedge \square(J=2\cdot I)$,
 Memory: $\langle\langle\rangle,\langle\rangle,\langle\rangle\rangle$.
After state 1:
 Program: $\circledcirc\big[empty \wedge (I=2) \wedge (I\ gets\ I+1) \wedge \square(J=2\cdot I)\big]$,
 Memory: $\langle\langle 1\rangle,\langle 2\rangle,\langle false\rangle\rangle$.

Before state 2:
 Program: $empty \wedge (I=2) \wedge (I\ gets\ I+1) \wedge \square(J=2\cdot I)$,
 Memory: $\langle\langle\rangle,\langle\rangle,\langle\rangle\rangle$.
After state 2:
 Program: $\circledcirc\big[false \wedge (I\ gets\ I+1) \wedge \square(J=2\cdot I)\big]$,
 Memory: $\langle\langle 2\rangle,\langle 4\rangle,\langle true\rangle\rangle$.

process. Furthermore, when a Tempura *request* or *display* statement is reduced, communication with the user takes place.

As noted in the presentation of Tempura's syntax, a formula such as

$$(I = 2) \wedge (J = I + 3)$$

can be viewed as both a statement and as a boolean test. In addition, a variable such as I can be considered either a location or an expression. Therefore, there exist two other reduction routines. The procedure *transform_loc* is for locations and the procedure *transform_expr* is for expressions. These are both used by *transform_stmt* and are described in sections 8.5 and 8.6, respectively.

Let us now consider the behavior of *transform_stmt* on individual types of Tempura statements.

8.4.1 Implementing the statements *true* and *false*

When the statement *true* is encountered, it is immediately reduced to the form ⓦ *true*. The values of the memory cells are not affected. Thus, *true* can be thought of as a no-operation statement. On the other hand, when *false* is encountered as a statement, the interpreter terminates execution with an error. This provides the user a way to abnormally stop the program if inconsistencies are detected.

8.4.2 Implementing equalities

An equality has the general form $l = e$, where l is a location and e is an expression. This is executed by first transforming l and e to new forms l' and e'. See sections 8.5 and 8.6 for details. If either l' or e' is not yet fully reduced, the statement

$$l' = e'$$

is returned as the result. If both are successfully reduced, then e' is a constant c and l' has one of the following two forms:

$$@loc(k), \quad \bigcirc l''.$$

The construct $@loc(k)$ is an internal location descriptor that references the memory cell indexed by the integer k. If l' is such a descriptor, the equality is processed by placing the constant c in the k-th cell. A check is made to ensure that the cell is empty prior to the assignment. The overall statement is then reduced to the form ⓦ $true$. This is returned as the result.

If the location l' is of the form $\bigcirc l''$ then the assignment is transformed to the conditional statement

$$if\ more\ then\ ⓦ(l'' = c)\ else\ false.$$

This postpones the actual assignment to the next state. The $more$ construct is used to ensure that the interval does indeed continue. If the test fails, the statement $false$ is executed, thus resulting in an error.

8.4.3 Implementing $empty$ and $more$

The transformation of the statement $empty$ places the value $true$ in the done-flag currently indexed by the variable $Current\_Done\_Cell$. The statement is then reduced to ⓦ $false$. The transformation of $more$ places the value $false$ in the done-flag indexed by $Current\_Done\_Cell$ and then reduces to ⓦ $true$. In the case of both statements, a check is made that the done-flag is empty (i.e., equals $\langle\rangle$) prior to the assignment.

8.4.4 Implementing $request$ and $display$

A statement of the form

$$request(l_1, \ldots, l_n)$$

is processed by first prompting the user for n values $c_1, \ldots, c_n$ and then transforming the statement to a conjunction of equalities

$$(l_1 = c_1) \wedge \cdots \wedge (l_n = c_n).$$

This is then re-reduced.

During the execution of a statement

$$display(e_1, \ldots, e_n),$$

the expressions $e_1, \ldots, e_n$ are all transformed. If any fail to completely reduce, the statement is returned unchanged. Otherwise, the resulting values are displayed and the statement is reduced to ⓦ $true$.

8.4.5 Implementing conjunctions

A conjunction $w_1 \wedge w_2$ is processed by first reducing both operands to forms w_1' and w_2'. If this is successful, the statement w_1' will be of the form $Ⓦ w_1''$ and the statement w_2' will be of the form $Ⓦ w_2''$. The overall statement can then be reduced to the form $Ⓦ(w_1'' \wedge w_2'')$. If either w_1'' or w_2'' is *true*, it can be omitted from the result. If w_1' or w_2' is not yet fully reduced, the conjunction $w_1' \wedge w_2'$ is returned instead.

8.4.6 Transforming $Ⓦ w$, $\bigcirc w$ and $\square w$

A statement of the form $Ⓦ w$ is already in reduced form and requires no further processing. Statements of the forms $\bigcirc w$ and $\square w$ are rewritten using the following equivalences and then reprocessed:

$$\bigcirc w \equiv \mathit{more} \wedge Ⓦ w$$
$$\square w \equiv w \wedge Ⓦ \square w.$$

8.4.7 Implementing implication

A conditional statement of the form $b \supset w$ is treated by first reducing the boolean expression b to either *true* or *false*. If b reduces to *true*, the overall statement is changed to w and again transformed. If b reduces to *false*, the result is $Ⓦ$ *true*.

8.4.8 Implementing some other statements

As was mentioned earlier, statements using the operators $\bigcirc$, *if*, *gets*, *stable* and *halt* can be converted to equivalent forms expressed in terms of the constructs described above. For example, a statement l *gets* e is first expanded to the form

$$\square\big[\mathit{more} \supset ([\bigcirc l] = e)\big]$$

and then re-reduced. Execution efficiency can be greatly increased if each new operator has its own specialized transformation sequence.

8.5 Implementing Locations

The procedure *transform_loc* has the form

$$\mathit{transform\_loc}(Loc).$$

The parameter *Loc* contains a location construct that is to be reduced. A location that is a variable such as I or J is transformed by converting it to an internal descriptor *@loc(k)*, where k is the variable's cell number as indexed by the environment contained in *Current_Env*. Locations can also be of the form $\bigcirc l$, where l is itself a location. These are considered to already be in reduced form.

8.6 Implementing Expressions

The procedure *transform_expr* has the form

transform_expr(Expr).

It attempts to reduce the contents of the parameter *Expr* using transformations that are suitable for expressions. An expression is in reduced form if it is either an arithmetic or boolean constant. Since constants such as 3 and *true* are already in reduced form, they require no further processing. A variable such as I is handled by first reducing it as a location. The result is an internal location descriptor of the form *@loc(k)*. The actual value of k is determined by looking at I's entry in the current environment. The descriptor is immediately re-reduced as an expression.

A location descriptor *@loc(k)* is itself transformed by examining the memory cell indexed by the integer k. If that cell is empty (i.e., equal to $\langle\rangle$), the descriptor is returned unchanged. Otherwise, the value stored in the cell is returned instead of the descriptor. For example, if the cell equals the singleton $\langle 4 \rangle$, the constant 4 is the result.

An expression such as the sum $e_1 + e_2$ is handled by first transforming e_1 and e_2 and then adding them if both are successfully reduced to constants. A conditional expression of the form *if b then e_1 else e_2* is transformed by reducing b and then selecting either e_1 or e_2 for further processing. Boolean expressions such as $\neg b$ and $b_1 \wedge b_2$ can be expanded to conditional expressions and then re-reduced. The boolean expressions *empty* and *more* are transformed to the constructs *@loc(done-cell)* and $\neg$*@loc(done-cell)*, respectively, where *done-cell* is the value of *Current_Done_Cell*. The new expressions are then immediately re-reduced.

8.7 Static Variables, Lists and Quantifiers

We now consider how to extend the interpreter to handle static variables, lists and quantifiers. The operators *len*, *fin* and ← as well as predicates and functions are also discussed.

8.7.1 Implementing static variables

Static variables are implemented by altering the memory slightly to include an extra boolean flag in every cell. This is set to *true* if the cell has been designated as static. When a memory cell is allocated for a state variable, the cell's flag is set to *false*. While clearing memory, the interpreter only empties those cells whose flag equals *false*. Thus once a value is put in a static cell, it is never lost.

8.7.2 Implementing lists

The value of a list is represented as a special list descriptor of the form

$$@list(length, offset).$$

Here *length* equals the number of list elements and *offset* is the index of the first of a series of consecutive memory cells for storing the elements.

8.7.3 Implementing *list*, *fixed_list* and *stable_struct*

A statement of the form $list(l, e)$ is processed by first reducing the location l and expression e. A series of consecutive memory cells are then allocated, one for each list element. If the cell referenced by location l is designated static, then so are these cells. Afterwards, l's cell is assigned a suitable list descriptor.

The implementation of the construct $fixed\_list(l, e)$ mentioned in section 7.3 is similar to that of $list(l, e)$. However, the memory cells initially allocated for the list elements continue to be used during the entire interval. This is done by simply reassigning the list description stored in location l from state to state. Thus a statement such as

$$fixed\_list(Table, 20)$$

is much more efficient that the logically equivalent form

$$\Box\ list(Table, 20)$$

since storage for the 20 elements of *Table* need only be allocated once rather than in every state. Furthermore, by using *fixed_list*, we ensure that the locations of the elements of *Table* do not change over time. This turns out to be very important when passing them as parameters to temporal predicates.

The construct *stable_struct l* mentioned in section 7.1.1 has an implementation similar to that for *fixed_list* in that the list descriptor stored in l is repeatedly reassigned throughout execution. However, the list elements of l are assumed to be already allocated. The conjunction

$$list(l, e) \wedge stable\_struct\ l$$

can in fact be used to implement the statement

$$fixed\_list(l, e).$$

8.7.4 Implementing list constructors

A list constructor of the form

$$\langle e_0, \ldots, e_{n-1} \rangle$$

is evaluated by first reducing the expressions $e_0, \ldots, e_{n-1}$ and then allocating n consecutive static memory cells in which the values of $e_0, \ldots, e_{n-1}$ are stored. A suitable list descriptor is then returned as the overall value of the expression. The iterative list constructor is implemented in a similar way.

8.7.5 Implementing subscripts

A subscripted location $l[e]$ is processed by first evaluating the location l and the expression e. The value of l's cell is fetched and checked to ensure that it is a list. Furthermore, a check is made to ensure that e's value is within the list's range. A location of the form $@loc(k)$ is then returned, where k is the sum of the list's offset into memory and e's value.

An expression $e_1[e_2]$ is treated in a similar manner. However once the relevant element's location has been determined, the value in the corresponding memory cell is fetched.

The sublist construct $l[e_1 .. e_2]$ mentioned in section 7.3 is implemented by first reducing the location l and the expressions e_1 and e_2. We then form a list descriptor with offset to the element in l referenced by e_1 and with length $e_2 - e_1$. This descriptor is stored in a new cell and the location of that cell is the result of the reduction.

8.7.6 Implementing the list-length operator

An expression $|e|$ is reduced by first determining the list descriptor of the expression e and then returning the length field stored in it.

8.7.7 Implementing existential quantification

A statement of the form

$$\exists V_1, \ldots, V_n : w$$

is processed by creating a new environment *env* containing entries for the quantified variables $V_1, \ldots, V_n$. Each of these variables has a fresh memory cell allocated for it. The entries in *env* for other variables are the same as in the surrounding environment contained in *Current_Env*. The statement is then transformed to an internal construct of the form

$$@exists(env, w).$$

This is immediately re-reduced. Such a statement is evaluated by saving the contents of *Current_Env*, setting *Current_Env* to *env* and then transforming the statement w. Afterwards, the old contents of *Current_Env* are restored. If the transformation result w' is successful, it has the form $\circledcirc\ w''$. Therefore, the overall statement is converted to the form

$$\circledcirc\ [\,@exists(env, w'')\,]$$

and returned. If w' is not fully reduced, the overall statement is given the form

$$@exists(env, w')$$

and returned.

8.7.8 Implementing universal quantification

A statement of the form

$$\forall v < e \colon w$$

can be readily handled by first reducing the expression e to a constant c and then building a conjunction of the form

$$\bigl(\exists v \colon [(v = 0) \wedge w]\bigr) \wedge \bigl(\exists v \colon [(v = 1) \wedge w]\bigr) \wedge \cdots \wedge \bigl(\exists v \colon [(v = c-1) \wedge w]\bigr).$$

The conjunction is immediately re-reduced.

8.7.9 Implementing the operators *len*, *fin* and ←

The implementation of a statement $len(e)$ first reduces the expression e to another expression e'. If this is not a constant, the overall statement is changed to $len(e')$ and returned. On the other hand, if e' is a constant c, the statement is reduced using the equivalence

$$len(c) \quad \equiv \quad \textit{if } (c = 0) \textit{ then empty else } \bigcirc len(c-1).$$

A statement of the form *fin w* is rewritten as the logically equivalent formula

$$\textit{if empty then } w \textit{ else } (\odot \textit{ fin } w)$$

and then immediately re-reduced. A temporal assignment $l \leftarrow e$ is processed by first reducing the expression e to a constant c. The overall statement is then changed to the form

$$fin(l = c).$$

This is immediately re-reduced.

8.7.10 Implementing predicate and function definitions

A predicate definition has the form

$$\textit{define } p(V_1, \ldots, V_n) \equiv w.$$

We execute this by creating a special descriptor of the form

$$@predicate(env, \langle V_1, \ldots, V_n \rangle, w),$$

where env is the current contents of the environment variable $Current\_Env$. The descriptor is stored in the predicate variable p's memory cell and provides enough information to properly access the predicate when it is invoked. The overall predicate definition is then reduced to the form $\circledcirc$ *true*. Function definitions are similarly handled.

8.7.11 Implementing predicate and function invocations
A predicate invocation has the form

$$p(e_1, \ldots, e_n).$$

We first access the value in p's memory cell and determine that it is indeed a special descriptor of the form

$$@predicate(env, \langle V_1, \ldots, V_n \rangle, w).$$

The locations of actual parameters $e_1, \ldots, e_n$ are then reduced. If any are in fact constants or expressions, their values are stored in freshly allocated static memory cells. A new environment env' is created in which each formal parameter $V_1, \ldots, V_n$ has an entry pointing to the corresponding actual parameter's location. The entries for other variables are made identical to those found in the predicate-descriptor's environment env.

We now transform the predicate invocation to the statement

$$@call(env', w),$$

where env' is the environment just constructed and w is the statement contained in the predicate-descriptor. The internal operator $@call$ is then processed by saving the current contents of the variable $Current\_Env$, setting $Current\_Env$ to env', reducing w and then restoring $Current\_Env$'s old value. If the transformed statement w' is completely reduced, then it has the form $\circledcirc w''$. The overall invocation is then rewritten as follows:

$$\circledcirc \left[@call(env', w'') \right].$$

On the other hand, if the transformation is not complete, the statement

$$@call(env', w')$$

is returned so that this can be executed during the current state's next pass.

Function invocations are similarly handled. They are eventually reduced to something of the form

$$@call(env, c),$$

where env is an environment and c is a constant. At this time, we discard the operator $@call$ and the environment and transform the overall expression to be simply the constant c.

8.8 Implementing *chop* and Iterative Operators

We now turn to the implementation of *chop* and related operators such as *for* and *while*.

8.8.1 Implementing the operator *chop*

The *chop* construct has the form

$$w_1; w_2,$$

where w_1 and w_2 are formulas. The first step in processing this is to allocate a memory cell to serve as the local done-flag associated with the subinterval in which w_1 is executed. The *chop* statement is then transformed to the internal form

$$@chop(done\text{-}cell, w_1, w_2).$$

Here *done-cell* is the index of the memory cell serving as the local done-flag.

The construct $@chop$ is executed by first saving the current value of the variable *Current_Done_Cell* and setting *Current_Done_Cell* to the index *done-cell*. The statement w_1 is transformed in this context. Afterwards, we restore the old value of *Current_Done_Cell*. If the transformation's result w' is not yet fully reduced, the $@chop$ statement is altered to have the form

$$@chop(done\text{-}cell, w', w_2)$$

and then returned as the overall result. However, if w' is fully reduced, it has the form $\circledcirc\, w''$. The *@chop* statement is therefore transformed to the following conditional form:

$$\textit{if } @loc(\textit{done-cell}) \textit{ then } w_2 \textit{ else } \bigcirc @chop(\textit{done-cell}, w'', w_2).$$

This is immediately re-reduced. Thus, if the value of the local done-flag is true, the subinterval in which w_1 was executed is empty and therefore w_2 can be started right away. If the local done-flag is false, then the execution of the *@chop* statement is continued to the next state of the overall interval. Note that the operator *strong-next* is used to indicate that the overall interval is not yet finished.

8.8.2 Implementing iterative operators
A statement of the form

for e times do w

is transformed by first reducing the expression e to a constant c. The overall statement is then rewritten as the conditional form

$$\textit{if } c = 0 \textit{ then empty else } \bigl(w; [\textit{for } c - 1 \textit{ times do } w]\bigr)$$

and again reduced.

The implementation of an indexed for-loop of the form

for $v < e$ do w

first reduces the expression e to a constant c. The statement is then transformed to the related construct

for $0 \leq v < c$ do w.

This is expanded using the following rule:

for $c_1 \leq v < c_2$ do w $\equiv$
  if $c_1 \geq c_2$ then empty
  else $\bigl([\exists v\colon (v = c_1 \wedge w)]; [\textit{for } c_1 + 1 \leq v < c_2 \textit{ do } w]\bigr).$

A while-loop is reduced using the equivalence

while b do w $\equiv$ (*if b then* [*w; while b do w*] *else empty*).

The iterative constructs *repeat* and *loop* are treated similarly.

8.8.3 Implementing the operator *skip*

The operator *skip* is first transformed to the form

$\bigcirc empty$

and then re-reduced.

8.9 Alternative Interpreters

The interpreter described here represents one technique for executing Tempura programs. It is rather easy to understand but suffers from being relatively slow. Let us now consider some alternative approaches and features. Most of them increase execution efficiency at the expense of generality. With a proper mix of these techniques, we feel that we can achieve speeds comparable with conventional imperative programming languages.

8.9.1 Immediate assignments

When the interpreter encounters the equality

$(\bigcirc I) = 1$

it postpones the actual assignment to I until the next state by using the form

$\text{\textcircled{w}}(I = 1).$

The interpreter can be modified to perform the assignment immediately. This leads to increased execution efficiency. However, one drawback is that statements with parallel assignments such as the following are not properly handled:

$([\bigcirc I] = I + 1) \wedge ([\bigcirc J] = J + I).$

In this example, we would alter I's value before being able to compute the expression $J + I$. Thus, the next value of J would be unknown. One way to get around this is to reorder the statement as follows:

$([\bigcirc J] = J + I) \wedge ([\bigcirc I] = I + 1).$

8.9.2 Two-level memory

So far, the interpreter we have presented maintains a single data value for each location. It is often attractive to maintain two such values: one for the current state and one for the next state. This permits the technique of immediate assignment described above to properly work on parallel assignments such as

$$([\bigcirc I] = I + 1) \wedge ([\bigcirc J] = J + I).$$

Note that static locations still only require a single value.

8.9.3 Single-pass processing

The current interpreter can make a number of left-to-right passes over a statement within each executed state. For example, the compound statement

$$(J = I + 2) \wedge (I = 1)$$

requires two passes. During the first pass the value for I is determined and during the second pass the value for J is determined. On the other hand, the following statement can be completely reduced in a single pass:

$$(I = 1) \wedge (J = I + 2).$$

If we use an interpreter having immediate assignment, then many useful programs can be written which require only one pass per executed state. Once they are debugged, such programs have the potential of being executed much faster since the interpreter need not do the extra processing required to detect and handle multiple passes. For example, when a memory cell is read, no test need be made regarding whether a value is already stored.

8.9.4 Time stamps

A time stamp can be included as part of each memory cell. Whenever the cell is stored into, the current state number is included with the data. When the cell is accessed, its state number can be checked to ensure that the data is current and not left over from a previous state. This approach eliminates the need to empty all cells at the beginning of every state. Many debugged programs require only one pass per executed state and can therefore be executed without the interpreter having to regularly empty memory cells or use time stamps.

8.9.5 Ignoring the operator *stable*

As we have noted, many debugged Tempura programs can be run using one pass per state and without emptying memory cells or maintaining time stamps. Any statement of the form *stable l* in such programs can be ignored since the value of the location l's memory cell will automatically remain unchanged. If l is a list, then the effect will safely propagate to its elements. Thus, the savings gained by not processing *stable* can be considerable. This technique should be used with great care since time-dependent errors can go undetected.

8.9.6 Redundant assignments

The interpreter does not permit two or more redundant assignments to the same location to occur in a single state. For example, the statement

$$(I = 1) \wedge (I = 1)$$

causes an error in processing even though it is logically equivalent to the statement

$$I = 1.$$

The principle of nonredundancy also applies to interval termination constructs such as *empty*. Only one such construct should be used in each state of an interval. Thus a formula such as

empty $\wedge$ *empty*,

although logically consistent, results in an error when executed. Tempura constructs such as the operators *empty*, *more*, *halt* and $\bigcirc$ affect interval length. Constructs such as $\wedge$, Ⓦ and $\square$ do not in themselves specify anything about termination.

If desired, the interpreter can be modified to permit the kind of redundant assignments mentioned here. We currently have reservations but more experience is needed to resolve this issue.

8.9.7 Special-purpose constructs

Various constructs can be added to Tempura in order to speed execution. For instance, the operator $:=$ is called *unit assignment* and is like temporal assignment ($\leftarrow$). However, unit assignment only works in intervals of length 1 and turns out to be more efficiently implementable. A statement $l := e$ is first transformed to the form

$$(\bigcirc l) = e$$

and then re-reduced. In addition, a test is made to ensure that the interval is indeed of unit length. Note that $:=$ can also be viewed as a restricted form of *gets*.

8.9.8 Suppressing checks

Programming systems usually have facilities for checking subscript ranges, detecting undefined variables and performing other such tests. The Tempura interpreter includes additional consistency checks regarding temporal behavior. For example, a statement such as

$$(I = 1) \wedge (I = 2)$$

is erroneous since the same variable cannot receive two values in one state. Similarly, the interpreter detects an error in the following program because no value is specified for the variable I in the second state:

$$(I = 1) \wedge \bigl[skip; (skip \wedge [I \leftarrow I + 1])\bigr] \wedge \square\ display(I).$$

These types of temporal checks can be suppressed. This increases the speed of execution but of course can result in various time-dependent bugs going unnoticed.

8.9.9 Call-by-name

When a function or predicate is invoked, the interpreter determines the locations of all actual parameters and uses them to build an environment. The effect is to implement parameter passing using call-by-reference. As in other programming languages, this is more efficient than call-by-name but has certain drawbacks. For example, when a subscripted element such as $L[I]$ occurs as a parameter, its location is determined only once. Therefore any change in the value of the index I is not reflected in the corresponding formal parameter. Our experience is that most subscripts tend to be static expressions (e.g., $L[1]$ or $L[i]$) so this is not a great limitation. A more subtle problem occurs if one specifies that L is a list using a statement such as

$\square$ $list(L, 3)$.

In every state, new memory cells are allocated for L's three elements. Therefore, a reference to, say, $L[2]$ will change over time. Thus, if $L[2]$ is passed as a parameter to a temporal predicate, the reference used becomes obsolete after the first state. This is one reason we use the constructs *fixed_list* and *stable_struct* for creating and maintaining such lists. Both of them ensure that the locations of list elements do not move around.

If one desires, it is not hard to alter the interpreter to implement parameter passing using call-by-name. It should be noted that the current interpreter permits one to simulate call-by-name by means of parameterless functions. For example, suppose we define the function f and the predicate p as follows:

$define\ f() = L[I],$
$define\ p(g) \equiv \bigl(g()\ gets\ [g() + 1]\bigr).$

The predicate invocation $p(f)$ is therefore equivalent to the statement

$L[I]\ gets\ (L[I] + 1).$

9 Experimental Features

In this section we examine some experimental constructs that are not especially well understood, yet have interesting applications and properties. The first involves the concept of temporal projection. Following this is a brief look at lambda expressions and their application to representing pointers. Finally, we discuss the use of the *process* construct in parallel programs and the use of the *prefix* construct in specifying the premature termination of computations.

9.1 Temporal Projection

When modelling hardware, it is natural to look at a circuit's behavior at different granularities of time. For example, the units of time might correspond to nanoseconds or clock ticks. We use the term *temporal projection* to denote the process of mapping from one level of time to another. In earlier work [14,34] we looked at one way to add operators for temporal projection to ITL. Since then we have developed an approach which is a bit easier to use and can be readily incorporated into Tempura. The main new construct is a formula of the form

$$w_1 \; proj \; w_2,$$

where w_1 and w_2 are themselves formulas. This has the following semantics:

$\mathcal{M}_\sigma[\![w_1 \; proj \; w_2]\!] = true$ iff
for some relative times $\tau_0, \tau_1, \ldots, \tau_m \in \{0, 1, \ldots, |\sigma|\}$,
the following are true:

$$0 = \tau_0 \leq \tau_1 \leq \cdots \leq \tau_m = |\sigma|,$$
and $\mathcal{M}_{\langle \sigma_{\tau_i} \ldots \sigma_{\tau_{i+1}} \rangle} [\![w_1]\!] = \textit{true}$, for all $i < m$,
and $\mathcal{M}_{\langle \sigma_{\tau_0} \sigma_{\tau_1} \ldots \sigma_{\tau_m} \rangle} [\![w_2]\!] = \textit{true}$.

This definition makes the formula w_1 *proj* w_2 true on any interval meeting two conditions. First, the interval can be broken up into a series of consecutive subintervals, each having the form $\sigma_{\tau_i} \ldots \sigma_{\tau_{i+1}}$ for some $i < m$ and satisfying the formula w_1. Second, the projected interval $\langle \sigma_{\tau_0} \sigma_{\tau_1} \ldots \sigma_{\tau_m} \rangle$ formed from the string of end-states of the subintervals satisfies w_2.

Consider, for example, the formula

$$\textit{len}(2) \ \textit{proj} \ \bigl[\textit{len}(4) \land (I = 0) \land (I \ \textit{gets} \ I + 1) \bigr].$$

This is true on any interval σ whose length is 8 and in which I's value starts at 0 and increases by 1 from σ_0 to σ_2, from σ_2 to σ_4 and so forth. To show this, we use the definition of *proj* with m equalling 4 and with τ's elements having the assignments

$$\tau_0 = 0, \quad \tau_1 = 2, \quad \tau_2 = 4, \quad \tau_3 = 6, \quad \tau_4 = 8.$$

Thus the left operand $\textit{len}(2)$ is true on each of the following subintervals of σ:

$$\langle \sigma_0 \sigma_1 \sigma_2 \rangle, \quad \langle \sigma_2 \sigma_3 \sigma_4 \rangle, \quad \langle \sigma_4 \sigma_5 \sigma_6 \rangle, \quad \langle \sigma_6 \sigma_7 \sigma_8 \rangle.$$

The right operand

$$\textit{len}(4) \land (I = 0) \land (I \ \textit{gets} \ I + 1)$$

is true on the projected interval

$$\sigma_0 \sigma_2 \sigma_4 \sigma_6 \sigma_8.$$

Figure 9.1 illustrates the projection pictorially. The value of I in odd-numbered states is not specified. This example is in fact logically equivalent to the formula

$$(I = 0) \land \textit{for 4 times do} \ (\textit{len}(2) \land [I \leftarrow I + 1]).$$

Note that different projections can be combined in parallel as well as with other kinds of formulas.

9.1.1 Incorporating *proj* in **Tempura**

We add projection to Tempura by permitting statements of the form

$$w_1 \ proj \ w_2.$$

Here w_1 and w_2 are themselves statements. Let us now look at two Tempura programs based on this construct.

Example (Describing intermediate states):

The following formula initializes the variable M to 1 and then doubles it in every third state for 4 times:

$$\bigl([len(2) \wedge stable\ M];\ skip\bigr) \\ proj\ \bigl[len(4) \wedge (M = 1) \wedge (M\ gets\ 2M)\bigr]. \tag{9.1}$$

Furthermore M remains stable during intermediate states of the projection. Figure 9.2 gives an execution depicting the behavior of M.

Example (Variable-length projection):

Let us define the predicate *count_and_sum* as follows:

$$count\_and\_sum(I, J) \quad \equiv_{\text{def}} \\ len(4) \wedge (I = 0) \wedge (J = 0) \wedge (I\ gets\ I+1) \wedge (J\ gets\ J+I).$$

This initializes the variables I and J to 0 and then repeatedly increases I by 1 and J by I for four units of time. The following

Figure 9.1: Example of projection

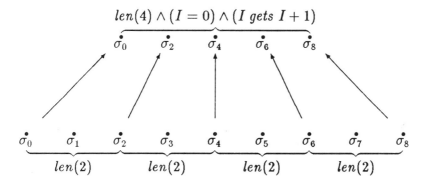

formula uses this predicate in a projection that has ever widening gaps dependent on I's values:

$$\bigl([len(I) \wedge stable\ I \wedge stable\ J]; skip\bigr) \\ proj\ [count\_and\_sum(I,J)]. \tag{9.2}$$

The projection has the values of I and J remain stable in intermediate states. This formula is executed in figure 9.3. Note that if we are not interested in the behavior of I and J in intermediate states, the projection can be specified as follows:

$$[len(I+1)]\ proj\ [count\_and\_sum(I,J)].$$

9.1.2 Universal projection

The kind of projection so far presented can be called *existential projection* since it is true on an interval if there is *at least one way* to break up the interval. The operator

$$w_1\ uproj\ w_2$$

is called *universal projection* and is true on an interval σ if for *every way* we break up σ using w_1, the projected result satisfies

Figure 9.2: Execution of formula (9.1)

```
State  0: M= 1
State  1: M= 1
State  2: M= 1
State  3: M= 2
State  4: M= 2
State  5: M= 2
State  6: M= 4
State  7: M= 4
State  8: M= 4
State  9: M= 8
State 10: M= 8
State 11: M= 8
State 12: M=16

Done! Computation length = 12.
```

w_2. We can express *uproj* as the dual of *proj*:

$$w_1 \text{ uproj } w_2 \quad \equiv_{\text{def}} \quad \neg(w_1 \text{ proj } \neg w_2).$$

At present, we do not see how to include universal projection in Tempura.

Example (Projection of a clocked system):

Universal projection provides a way of abstracting from digital behavior involving an explicit clock signal to behavior at the register-transfer level. As an example of this, let us consider a simple system driven by clock pulses. We first define the pulse operator $\uparrow\downarrow \textit{Clock}$ to be true on intervals where the bit signal *Clock* rises and then falls:

$$\uparrow\downarrow \textit{Clock} \quad \equiv_{\text{def}} \\ (\textit{Clock} \approx 0); \textit{skip}; (\textit{Clock} \approx 1); \textit{skip}; (\textit{Clock} \approx 0).$$

The system we have in mind has the three bit signals *Clock*, X and Y. Universal projection facilitates examining those properties of X and Y that are true across clock pulses independent of where we mark the beginning and end of each individual pulse. We can use projection of the form

$$(\uparrow\downarrow \textit{Clock}) \text{ uproj } w$$

Figure 9.3: Execution of formula (9.2)

```
State  0: I= 0  J= 0
State  1: I= 1  J= 0
State  2: I= 1  J= 0
State  3: I= 2  J= 1
State  4: I= 2  J= 1
State  5: I= 2  J= 1
State  6: I= 3  J= 3
State  7: I= 3  J= 3
State  8: I= 3  J= 3
State  9: I= 3  J= 3
State 10: I= 4  J= 6

Done!  Computation length = 10.
```

to specify and reason about a formula w describing the behavior of X and Y at the register-transfer level. For example, the following formula is true if X is repeatedly inverted and Y remains stable over a series of clock pulses:

$$(\uparrow\downarrow Clock)\ uproj\ \big[(X\ gets\ \neg X) \wedge stable\ Y\big].$$

9.2   Lambda Expressions and Pointers

Lambda expressions provide a natural means of representing functions as values. We now briefly sketch how to incorporate them into ITL and Tempura. We also show how lambda expressions can be used to represent pointers.

9.2.1   Lambda expressions

A lambda expression has the form

$$\lambda V_1,\ldots,V_n\!:e,$$

where V_1, ..., V_n are variables and e is an expression. The body of e can itself contain lambda expressions. In addition, we permit lambda predicates having the syntax

$$\Lambda V_1,\ldots,V_n\!:w,$$

where V_1, ..., V_n are variables and w is a formula.

First order lambda constructs can be viewed as temporal functions and predicates. To date, we do not have an adequate semantics for higher order lambda expressions.

An invocation of a lambda construct has the form

$$e_0(e_1,\ldots,e_n),$$

where $n \geq 0$ and e_0, ..., e_n are expressions. The value of e_0 should be a lambda construct of arity n.

9.2.2   Pointers

We can represent a pointer to a variable A by means of the parameterless lambda expression

$$\lambda\!:A.$$

For example, suppose the static variable b has this as its value. The variable A can then be indirectly accessed through the function invocation $b()$. Thus, the following formula is true on intervals where A increases by 1:

$$[b = (\lambda : A)] \wedge [b() \leftarrow b() + 1].$$

Let us now introduce some operators to make the pointer notation look more conventional. The expression

$$ref(l)$$

equals a pointer to the location l. It can be defined as follows:

$$ref(l) \quad =_{\text{def}} \quad \lambda : l.$$

A pointer expression e can be dereferenced using the construct

$$deref(e).$$

It is defined as follows:

$$deref(e) \quad =_{\text{def}} \quad e().$$

Thus the sample formula given above can be also expressed as follows:

$$[b = ref(A)] \wedge [deref(b) \leftarrow deref(b) + 1].$$

9.3 The *process* Construct

In section 8.9.6, we mentioned that two separate statements can not in parallel determine the length of an interval. If one has a number of statements $w_1, w_2, \ldots, w_n$ to be run in parallel and each individually determines interval length, then all but one should be used with the unary operator *process*:

$$w_1 \wedge (process\ w_2) \wedge \cdots \wedge (process\ w_n).$$

For instance, the following statement is permitted:

$$\begin{aligned}&[(I = 0) \wedge (I\ gets\ I + 1) \wedge halt(I = 5)]\\&\wedge\ process\ [(J = 0) \wedge (J\ gets\ J + 2) \wedge halt(J = 10)].\end{aligned}$$

The predicate *sum_tree_process* in section 7.1.2 and the predicate *sort_process* in section 7.3.3 both make use of *process*. This is so that multiple instances of them can be run in parallel. Note that the *process* construct has no logical semantics and can be defined as follows:

$$process\ w\ \equiv_{def}\ w.$$

Its purpose is merely to explicitly indicate to the interpreter that redundant specifications of interval termination are present.

9.4 The *prefix* Construct

When a program is designed, it is often necessary to provide facilities for terminating execution before the normal end is reached. For example, an interpreter might detect an error in the code it is processing and therefore wish to leave a number of recursive calls via an error exit. In conventional Algol-like languages this kind of behavior is readily achieved by means of a go-to statement which can immediately exit from any level of nesting in a program. This technique appears inappropriate for Tempura since Tempura does not have go-to statements and the underlying ITL formalism seems incapable of supporting them.

It turns out that one can achieve the effect of error exits by means of the *prefix* operator. The formula *prefix w* is true on an interval σ iff σ is a prefix of some interval σ' on which w is itself true:

$$\mathcal{M}_\sigma[\![prefix\ w]\!] = true \quad \text{iff}$$
there is some σ' in $\mathcal{I}$ such that
$$|\sigma'| \geq |\sigma|,\ \mathcal{M}_{\sigma'}[\![w]\!] = true\ \text{and}\ \sigma = \langle \sigma'_0 \sigma'_1 \ldots \sigma'_{|\sigma|}\rangle.$$

For example, the following formula is true on any interval having length not greater than 5:

$$prefix\,[len(5)].$$

9.4.1 Incorporating *prefix* in Tempura

We introduce Tempura statements of the form *prefix w* where w is itself a statement. Here are some simple applications.

Example (Early termination of iteration):
Consider the following formula:

$$halt(I = 16) \land prefix\big[len(10) \land (I = 1) \land (I \text{ gets } 2I)\big]. \quad (9.3)$$

The operand of *prefix* specifies that the variable I repeatedly doubles over an interval of length 10. However, the outer *halt* construct overrides this and terminates the interval upon I reaching the value 16, i.e., after four units of time. Figure 9.4 shows the behavior of the formula.

Example (Early termination of nested recursion):
Figure 9.5 depicts a modified version of the serial tree summation program described earlier in section 7.1.1. The predicates reference a global variable *Tag* which normally equals the empty list $\langle\rangle$ except at the end of the execution of *pfx_sum_tree* when it is assigned the singleton $\langle true\rangle$. However, if an invocation of the predicate *pfx_sum_tree_body* finds a leaf having the value 0, the value of *Tag* is set to the singleton $\langle false\rangle$. The following program handles initialization and includes the *prefix* construct to facilitate early termination upon the detection of such a leaf:

$$\begin{array}{c}(Tree = initial\_tree) \land (Tag = \langle\rangle) \\ \land\; halt(|Tag| = 1) \land prefix\big[pfx\_sum\_tree(Tree)\big]. \end{array} \quad (9.4)$$

This invokes the predicate *pfx_sum_tree* and terminates when the variable *Tag* becomes a singleton. We assume that the static variable *initial_tree* equals the starting value for the variable *Tree*. Figure 9.6 shows the formula's behavior for two possible trees.

Figure 9.4: Execution of formula (9.3)

```
State   0:  I= 1
State   1:  I= 2
State   2:  I= 4
State   3:  I= 8
State   4:  I=16

Done!  Computation length = 4.
```

Figure 9.5: Tree summation with prefix computations

$define\ pfx\_sum\_tree(Tree) \equiv$
$pfx\_sum\_tree\_body(Tree);$
$($
$\quad skip \wedge (Tag \leftarrow \langle true \rangle) \wedge (stable\ Tree)$
$)$

$define\ pfx\_sum\_tree\_body(Tree) \equiv$
$if\ is\_integer(Tree)\ then\ ($
$\quad if\ Tree > 0\ then\ empty$
$\quad else\ ($
$\quad\quad skip \wedge (Tag \leftarrow \langle false \rangle) \wedge (stable\ Tree)$
$\quad)$
$)$
$else\ ($
$\quad pfx\_sum\_subtree(Tree, 0);$
$\quad pfx\_sum\_subtree(Tree, 1);$
$\quad ($
$\quad\quad skip \wedge (Tree \leftarrow Tree_0 + Tree_1) \wedge (stable\ Tag)$
$\quad)$
$).$

$define\ pfx\_sum\_subtree(Tree, i) \equiv$
$pfx\_sum\_tree\_body(Tree_i)$
$\wedge\ stable\_struct\ Tree$
$\wedge\ stable\ Tree_{1-i}.$

Figure 9.6: Execution of formula (9.4)

```
State 0: Tag=<>        Tree=<<<1,1>,<1,1>>,<<1,1>,<1,1>>>
State 1: Tag=<>        Tree=<<2,<1,1>>,<<1,1>,<1,1>>>
State 2: Tag=<>        Tree=<<2,2>,<<1,1>,<1,1>>>
State 3: Tag=<>        Tree=<4,<<1,1>,<1,1>>>
State 4: Tag=<>        Tree=<4,<2,<1,1>>>
State 5: Tag=<>        Tree=<4,<2,2>>
State 6: Tag=<>        Tree=<4,4>
State 7: Tag=<>        Tree=8
State 8: Tag=<true>    Tree=8

Done! Computation length = 8.


State 0: Tag=<>        Tree=<<<1,1>,<1,1>>,<<0,1>,<1,1>>>
State 1: Tag=<>        Tree=<<2,<1,1>>,<<0,1>,<1,1>>>
State 2: Tag=<>        Tree=<<2,2>,<<0,1>,<1,1>>>
State 3: Tag=<>        Tree=<4,<<0,1>,<1,1>>>
State 4: Tag=<false>   Tree=<4,<<0,1>,<1,1>>>

Done! Computation length = 4.
```

The first computation processes a tree containing all 1's. The second computation operates on a variant of the tree in which one leaf has the value 0.

9.5 Implementing these Constructs

We now discuss how the various constructs just introduced can be implemented in the Tempura interpreter. This presentation can be skipped if desired.

9.5.1 Implementing temporal projection

The interpreter handles a statement of the form w_1 *proj* w_2 by first allocating a new memory cell and then transforming the statement to the internal construct

$@proj(\textit{done-cell}, w_1, w_2).$

This is immediately re-reduced. Here *done-cell* is the index of the memory cell. The cell serves as a local done-flag for the projected interval in which the statement w_2 is executed.

We execute the *@proj* construct by first saving the value of the variable *Current_Done_Cell* and setting it to the index *done-cell*. The statement w_2 is then transformed in this context to a new statement w'. Afterwards the old value of *Current_Done_Cell* is restored. If w' is not yet fully reduced, the overall *@proj* statement is rewritten as

$@proj(\textit{done-cell}, w_1, w').$

This is returned as the result of the transformation. On the other hand, if w' is fully reduced and thus of the form $Ⓥ\, w''$, then the overall projection is transformed to the following conditional statement and then immediately re-reduced:

if $@loc(\textit{done-cell})$ *then empty*
else $[w_1;\, @proj(\textit{done\_cell}, w_1, w'')].$

This tests the localized done-flag indexed by *done-cell*. If it is true, the interval in which w_2 was transformed is finished and therefore the overall projection terminates. Otherwise, the statement w_1 is executed followed by the resumption of the projection statement.

9.5.2 Implementing lambda expressions and pointers

The lambda constructs implemented in the Tempura interpreter can uniformly handle both first-order and higher-order variants. The approach taken is basically the same as for processing predicate and function definitions. See section 8.7.10 for a presentation of this.

The technique of using lambda expressions for representing pointers actually works in the current Tempura interpreter. However, the constructs *ref* and *deref* can be implemented in a more efficient but less general manner by means of a new type of descriptor specifically for pointers. We omit the details.

9.5.3 Implementing the *process* construct

When reducing a statement of the form

process w

we first allocate a memory cell to serve as a local done-flag. The statement is then transformed to the internal construct

@*process*(*done-cell*, *w*)

and immediately re-reduced. Here *done-cell* is the index of the memory cell.

The @*process* construct is itself implemented by saving the value of the variable *Current_Done_Cell* and setting it to the index *done-cell*. The statement w is transformed within this context to a new statement w'. Afterwards the variable *Current_Done_Cell* is restored to its old value.

If the statement w' is not yet fully reduced, the statement

@*process*(*done-cell*, w')

is returned as the result. Otherwise, if w' is fully reduced then it has the form $\odot\, w''$ for some w''. We therefore transform the @*process* construct to the following conditional statement:

if [*empty* ≡ @*loc*(*done-cell*)]
then [$\odot$ @*process*(*done-cell*, w'')] *else false*.

This tests to make sure that the current done-flag and the local done-flag used by the *@process* operator agree in value. If they do, the statement w'' is placed within the *@process* construct in preparation for any subsequent states. If the done-flags are not equal, the statement *false* is executed, thus generating an error. Note that once a Tempura program is debugged, the consistency check between the done-flags can be suppressed.

9.5.4 Implementing the *prefix* construct

The *prefix* construct is implemented in the same way as the *process* construct. However, the following conditional statement is used in place of the one given previously:

$$\textit{if } \neg[\textit{more} \wedge @loc(\textit{done-cell})]$$
$$\textit{then } [\widehat{w} \; @prefix(\textit{done-cell}, w'')] \textit{ else false}.$$

The conditional test ensures that the interval of the prefixed statement does not terminate before the outer interval does.

10 Discussion

We now discuss the status of Tempura and some directions for further research. Afterwards we look at programming formalisms that seem related to Tempura and also review some other work on temporal logic.

10.1 Experience and Further Work

Using the ideas discussed here, we have implemented a prototype Tempura interpreter in Lisp. Its design is based on the interpreter presented in chapter 8 and it includes facilities for experimenting with some of the alternative execution strategies mentioned in section 8.9. A great variety of Tempura programs have been written and successfully run. Roger Hale, a PhD student at Cambridge University, has more recently implemented a faster version of the interpreter in the programming language C. In [13], he describes the application of ITL and Tempura to the modelling of a digital ring network. Another interpreter has been developed in Prolog by Masahiro Fujita, Shinji Kono and others at Tokyo University.

In the future we plan to build a compiler and bootstrap Tempura in itself. We also hope to describe the operational semantics of Tempura in ITL. This will enable us to formalize the relation between various execution strategies, both sequential and parallel. It seems likely that we will generalize ITL to permit infinite intervals in order to handle nonterminating computations.

So far we have made little mention of ITL's proof theory. This is much less developed than the model theory. Work by Halpern and Moszkowski has shown that one propositional subset of ITL is undecidable and that another is decidable (see Moszkowski [34] for details). Although it is not hard to come up with sound axioms and inference rules for ITL, no systematic work has been done to date. We therefore feel that ITL's proof theory represents a promising area for future research.

10.2 Related Programming Formalisms

Let us now consider some behavioral formalisms and programming languages that seem related to ITL and Tempura.

10.2.1 The programming language *Lucid*

The functional programming language *Lucid* [2,45] developed by Ashcroft and Wadge is similar to parts of Tempura. For example, the Lucid program

$$I = 0 \; fby \; (I+1); \quad J = 0 \; fby \; (J+I)$$

roughly corresponds to the temporal formula

$$(I = 0) \wedge (J = 0) \wedge (I \; gets \; I+1) \wedge (J \; gets \; J+I).$$

This illustrates how the operator *gets* can be handled in Lucid. On the other hand, Algol-like temporal constructs such as ←, *chop* and *while* do not have direct analogs in Lucid. Thus, a Tempura statement such as

$$while \; (M \neq 0) \; do \; \bigl(skip \wedge [M \leftarrow M - 1] \wedge [N \leftarrow 2N]\bigr)$$

cannot be readily translated. In [1], Ashcroft and Wadge develop a calculus for reasoning about Lucid programs.

10.2.2 *CCS* and *CSP*

Milner's *Calculus of Communicating Systems* [32] as well as Hoare's *Communicating Sequential Processes* [22] are popular notations for describing and reasoning about parallel systems. Related work includes Milne's *CIRCAL* [31] for modelling hardware and the CSP-inspired programming language *occam^{TM}* [24]. In CCS and CSP, multiple processes interact with one another by mutually synchronizing on events. There are operators for composing processes, waiting for events and concealing events.

Let us look at some of the constructs used in CSP. The form $a \to P$ denotes a process that awaits the event a and then executes the subprocess P. Similarly, the form

$$a \to (b \to P)$$

describes a process that first waits for the event a and then waits for b before executing the subprocess P. The construct $P \parallel Q$ runs the processes P and Q in parallel. Thus the form

$$(a \to P) \parallel (a \to Q)$$

has the event a separately trigger each of the processes P and Q. This is viewed as equivalent to having a trigger P and Q together:

$$(a \to P) \parallel (a \to Q) \quad = \quad a \to (P \parallel Q).$$

The CSP notation includes channels for synchronized communication between parallel processes. An event of the form $c!v$ sends the value of the variable v to the channel c. Similarly, the event $c?v$ awaits the receipt of an input from channel c and places the value in the variable v.

The treatment of time is not a central issue in CCS and CSP. However, it is possible to describe a clock process that serves as a source of events representing ticks. A variant of CCS called *Synchronous CCS* models concurrent systems that operate in lock step.

10.2.3 Predicative programming

Hehner [17] views programs as logical predicates that describe the input-output behavior of variables. Various Algol-based constructs such as assignment ("$:=$"), sequencing ("$;$") and while-loops are treated. Their semantics are given by means of special temporal operators. The construct $\grave{x}$ (read "*x in*") represents the value of the variable x before executing some statement. The analogous construct $\acute{x}$ (read "*x out*") represents the value of the variable x after the statement finishes. For example, the following formula specifies that x increases by 1 and y remains unchanged:

$$(\acute{x} = \grave{x} + 1) \wedge (\acute{y} = \grave{y}). \tag{10.1}$$

This is similar to the ITL formula

$$(X \leftarrow X + 1) \wedge (Y \leftarrow Y).$$

Hehner reduces the meaning of a statement to a formula based on these operators. For example, in a program with two variables x and y, the meaning of the simple assignment $x := x + 1$ might correspond to formula 10.1.

The precise amount of time taken by a statement can not be directly specified. However, if one desires to formalize properties regarding computation length, an extra clock variable can be used. Hehner then goes on to introduce concurrency with interprocess communication through CSP-like channels.

10.2.4 The programming language *Esterel*

Most programming languages have no formal notion of time. For example, even if such languages permit statements specifying delay, the semantics of these kinds of constructs are usually imprecise. Tempura is an exception to this rule as is the language Esterel presented by Berry and Cosserat [4]. Programs in Esterel can include constraints involving computation length. In figure 10.1 we show an example in which the duration of loops is specified in units of time or distance covered. Berry and Cosserat characterize the behavior of Esterel programs using transition rules based on a discrete model of time.

Figure 10.1: Sample Esterel program

```
var SPEED : int in
loop
    every 10 seconds do
        SPEED := 0;
        every METER do
            SPEED := SPEED + 1
        end
    end;
    emit SPEED_MEASURE(SPEED)
end
end
```

10.2.5 The programming language *Prolog*

The programming language *Prolog* [9,23] uses Horn clauses in first order logic as a means of describing computations. Kowalski [25,26], Colmerauer et al. [10] and others originally applied this approach to expressing algorithms for such tasks as natural language understanding and theorem proving. This led to the use of resolution as the basis for executing Prolog programs. Subsequent work by Clark, McCabe and Gregory [7,8], Shapiro [42], and others has generalized logic programming to deal with concurrent systems.

Let us consider how to express the following two predicate definitions in Prolog:

$$is\_double(I, J) \equiv_{\text{def}} (J = 2I),$$
$$is\_double1(I, J) \equiv_{\text{def}}$$
$$\text{if } I = 0 \text{ then } J = 0$$
$$\text{else } \exists I1, J1 : \big[(I1 = I - 1)$$
$$\wedge\, is\_double1(I1, J1) \wedge (J = J1 + 2)\big].$$

We assume that all the variables mentioned here range over the nonnegative integers. The definition of *is_double1* is recursive but well founded since the first parameter repeatedly decreases by 1 until it reaches 0. Both predicates are true if the second parameter J equals twice the first parameter I. Thus, the predicates are in fact equivalent:

$$\models \quad is\_double(I, J) \equiv is\_double1(I, J).$$

In figure 10.2 we show Prolog programs corresponding to both *is_double* and *is_double1*. The Prolog versions of the predicates illustrate how definitions generally consist of lists of implications. Furthermore, hidden variables are not explicitly quantified.

The developers of Prolog strongly believe in distinguishing between logic and any procedural aspects of implementation. As a consequence, Prolog and its offspring do not have any notion of time and therefore are unable to directly express imperative constructs such as assignments. Furthermore, there is no direct means of logically specifying or reasoning about such things as computation length and invariants.

In practice, extra-logical constructs such as *assert* and *retract* can be used to get around the lack of assignment statements. They provide a means for adding and removing facts from a system-maintained database. However, their usage is generally not considered good programming style. Alternatively, one can stay completely within the framework of the underlying logic by explicitly representing dynamically changing objects as lists of values. Whether this represents a practical and desirable way to express programs and properties remains to be seen.

Compared with Prolog, our approach does not limit itself to a subset of conventional logic based on Horn clauses. Nonetheless, it is a form of logic programming, albeit with temporal logic as the underlying formalism. We feel this provides a more natural setting for both programming and reasoning about dynamic systems. For example, here are two ITL definitions corresponding to procedural interpretations of the original logical predicates:

$$double(I, J) \equiv_{def} (empty \wedge [J = 2I]),$$
$$double1(I, J) \equiv_{def}$$
$$(J = 0)$$
$$\wedge \text{ while } I > 0 \text{ do } (skip \wedge [I \leftarrow I - 1] \wedge [J \leftarrow J + 2]).$$

These are readily executable in Tempura. Within ITL we can specify various time-dependent properties. For instance, in both algorithms the final value of J equals twice the initial value of I:

$$\models \quad double(I, J) \supset J \leftarrow 2I,$$
$$\models \quad double1(I, J) \supset J \leftarrow 2I.$$

In addition, the following property states that the length of a computation satisfying *double1* equals the initial value of I and

Figure 10.2: Prolog predicates for doubling a value

$$is\_double(I, J) \subset J \text{ is } 2 * I.$$
$$is\_double1(0, 0).$$
$$is\_double1(I, J) \subset$$
$$(I \neq 0) \wedge (I1 \text{ is } I - 1)$$
$$\wedge \ is\_double1(I1, J1) \wedge (J \text{ is } J1 + 2).$$

that during the computation, the value of the expression $2I + J$ remains stable:

$$\models \quad double1\,(I,J) \supset \bigl[len(I) \wedge stable(2I + J)\bigr].$$

Note that Tempura does not completely exclude the style of programming used in Prolog. For example, the original logical predicates *is_double* and *is_double1* can be directly embedded in Tempura programs such as the following:

$$len(5) \wedge (I = 0) \wedge (I \text{ gets } [I + 1])$$
$$\wedge \; \Box \; is\_double1\,(I,J) \wedge \Box \; display(I,J).$$

One significant difference is the lack of resolution and backtracking in Tempura. Perhaps a variant of Tempura can be designed that incorporates these features.

10.2.6 Functional programming

Functional programming [18] is based on the idea that certain types of functions can be interpreted as executable descriptions of computations. McCarthy's programming language *Lisp* [27] is perhaps the best known example of this approach. The functions themselves have no built-in notion of time. Therefore, as is the case with logic programs, dynamic behavior is in effect modelled indirectly. Recursion seems to be the most common technique used for this. Another approach is to represent time-dependent variables indirectly as lists or as functions containing an explicit time parameter.

Here is a simple recursive function for doubling a nonnegative value:

$$double\_func(I) \quad =_{\text{def}}$$
$$if\;(I = 0)\;then\;0\;else\;\bigl[2 + double\_func(I - 1)\bigr].$$

This has the following correctness property:

$$\models \quad double\_func(I) = 2I.$$

For reasons of efficiency, *tail recursion* is often used when implementing such functions. Here is a variant of *double_func* that is defined in this way:

$$double\_func1(I) =_{\text{def}} aux\_double\_func1(I, 0),$$
$$aux\_double\_func1(I, J) =_{\text{def}}$$
$$\text{if } (I = 0) \text{ then } J \text{ else } aux\_double\_func1(I - 1, J + 2).$$

The function *double_func1* is logically equivalent to *double_func*. The auxiliary function *aux_double_func1* satisfies the following property:

$$\models aux\_double\_func1(I, J) = 2I + J.$$

Note that all these definitions and properties make perfect sense in ITL. However, the iterative flow of control suggested by tail recursion can be directly expressed through while-loops and other such formulas.

Functional languages sometimes include constructs such as Lisp's *prog*, *setq* and *rplaca* in an ad hoc manner in order to permit in-place assignments and other imperative operations. Language purists tend to discourage their use. Even so, such features often seem indispensable for reasons of clarity and efficiency. For instance, suppose we are maintaining a 1000-element list and wish to periodically alter various elements. It is a great waste of space to have to create a completely new list on each occasion. Indeed, it is conceptually proper to view the various operations as being applied to a single, dynamically changing data structure. This is not directly possible in the functional framework.

Perhaps the biggest justification of functional programming has been its mathematical elegance and simplicity relative to conventional imperative languages. We feel that ITL and Tempura may offer an attractive middle ground which permits one to directly program in a procedural manner without compromising on formal rigor and without distancing oneself from the underlying implementation on computers with alterable memory.

10.3 Other Work on Temporal Logic

Let us now look at some work on using various kinds of temporal logics to specify and reason about dynamic systems.

10.3.1 Interval logic

Schwartz, Melliar-Smith and Vogt [41] develop a formalism called *interval logic* which includes temporal formulas having the syntax

$$[I]w.$$

Here I can be built from a variety of special constructs for indicating the scope of the interval in which the temporal formula w is to be evaluated. Thus I can be thought of as an "interval designator." For example, the formula

$$[(X = Y) \Rightarrow (Y = 16)] \,\Box\, (X > Z)$$

is true on an interval if X is greater than Z throughout the subinterval starting the first time X equals Y and ending when Y equals 16. Note that the designator is not itself a temporal formula. This is unlike the approach of ITL in which constructs used to specify subintervals are themselves always formulas.

10.3.2 Generalized *next* operator

Shasha, Pnueli and Ewald [43] propose some generalized forms of the *next* operator in which explicit time offsets are mentioned. For example, the formula

$$\bigcirc^e w$$

is true if the subformula w is true in e units of time from now. This is basically the same as the ITL formula

$$len(e); w.$$

The following two types of formulas are also permitted:

$$[e_1, e_2]w, \quad \langle e_1, e_2 \rangle w.$$

They are defined as follows:

$$[e_1, e_2]w \equiv_{def} \forall i\colon \bigl((e_1 \le i < e_2) \supset \bigcirc^i w\bigr),$$
$$\langle e_1, e_2 \rangle w \equiv_{def} \exists i\colon \bigl((e_1 \le i < e_2) \wedge \bigcirc^i w\bigr).$$

These constructs let one express behavior over various suffix subintervals of time but do not provide access to prefix subintervals in the way ITL's *chop* does. For example, the following ITL formula does not seems to be as elegantly expressed in their notation:

$$(K+1 \to K); (K+2 \to K).$$

10.3.3 Temporal logic as an intermediate language

Tang [44] uses temporal logic as the basis for a programming language called *XYZ/E*. Programs consist of a conjunction of transitions. An individual transition describes changes to be made to program variables and a special program counter. The temporal operators seem limited to $\bigcirc$ and $\square$. Figure 10.3 shows an example of this style. Note that although the operator ";" occurs in the program, the associated semantics seem to be those of logical-and ($\wedge$) rather than of *chop*. Tang includes some transformations that permit one to rewrite an Algol-like program in XYZ/E. This provides a way for giving temporal semantics to conventional programming constructs.

10.3.4 Semantics based on transition graphs

Manna and Pnueli [29] discuss ways of translating conventional programming constructs into transition systems described in temporal logic. The resulting temporal descriptions are then

Figure 10.3: Sample program in XYZ/E

$$\square \big[(\#lb = gcd) \Rightarrow (\bigcirc \#lb = l1);$$
$$(\#lb = l1) \Rightarrow (\bigcirc \#Ix = Ia) \wedge (\bigcirc \#lb = l2);$$
$$(\#lb = l2) \Rightarrow (\bigcirc \#Iy = Ib) \wedge (\bigcirc \#lb = l3);$$
$$(\#lb = l3) \wedge (\#Ix = \#Iy) \Rightarrow (\bigcirc \#lb = l6);$$
$$(\#lb = l3) \wedge (\#Ix \neq \#Iy) \Rightarrow (\bigcirc \#lb = l4);$$
$$(\#lb = l4) \wedge (\#Ix > \#Iy) \Rightarrow$$
$$\quad (\bigcirc \#Ix = \#Ix - \#Iy) \wedge (\bigcirc \#lb = l5);$$
$$(\#lb = l4) \wedge (\#Ix \leq \#Iy) \Rightarrow$$
$$\quad (\bigcirc \#Ix = \#Iy - \#Ix) \wedge (\bigcirc \#lb = l5);$$
$$(\#lb = l5) \Rightarrow (\bigcirc \#lb = l3);$$
$$(\#lb = l6) \Rightarrow (\bigcirc \#Iz = \#Ix) \wedge (\bigcirc \#lb = l7);$$
$$(\#lb = l7) \Rightarrow (\bigcirc \#lb = stop) \big]$$

used to reason about the original programs. The programs can include multiple processes which are executed through interleaving.

Figure 10.4 shows a sample program based on this approach. The program determines twice the value of the variable I and places the result in the variable J. Each statement is associated with a temporal formula characterizing the statement's behavior. For example, the effect of the assignment statement at location l_3 can be represented as follows:

$$\Box\bigl[at\ l_3 \supset \forall a,b\colon \bigl(\langle I,J\rangle = \langle a,b\rangle \supset \Diamond[at\ l_4 \wedge \langle I,J\rangle = \langle a-1,b\rangle]\bigr)\bigr].$$

Thus, whenever the program is at location l_3 it ultimately transfers to location l_4. In addition the value of I is decremented by 1 and the value of J remains unchanged. Invariants, termination properties and other issues can be dealt with. One drawback of this approach is its lack of compositionality since it can only deal with complete programs.

10.3.5 Compositional proof rules

Barringer, Kuiper and Pnueli [3] use a modified form of interval temporal logic as part of a compositional proof system for concurrent programs. Assertions have the syntax $\{S\}w$ where S is a statement and w is a temporal formula. For example the following proof rule describes the semantics of an assignment statement within an individual process:

$$\{v := e\}\bigl[(E)\,\mathcal{U}\bigl(\Pi \wedge (\bigcirc \bar{y} = \bar{y} \circ [v \leftarrow e])\bigr) \wedge \bigcirc(E\,\mathcal{U}\,fin)\bigr].$$

Figure 10.4: Algorithm for doubling a value

l_0: $I := n$
l_1: $J := 0$
l_2: *if* $I = 0$ *then goto* l_6
l_3: $I := I - 1$
l_4: $J := J + 2$
l_5: *goto* l_2
l_6: *halt*.

Here the variable $\bar{y}$ is a vector that associates values with the variables used by the process. The special proposition E is true if the process is inactive and Π is conversely true if the process is active. The *until* operator $\mathcal{U}$ is a temporal construct used to specify that the process remains inactive until the value of v in $\bar{y}$ is altered to equal e. In general, a temporal formula of the form $w_1\,\mathcal{U}\,w_2$ is true if the formula w_1 remains true until some time when formula w_2 is true. After performing the assignment, the process stays inactive for the rest of the interval.

The proof rule for the sequential composition of two statements S_1 and S_2 requires one to first demonstrate the assertions $\{S_1\}w_1$ and $\{S_2\}w_2$ for some formulas w_1 and w_2. From this one can immediately deduce the assertion

$$\{S_1; S_2\}(w_1\,\mathcal{C}\,w_2),$$

where $\mathcal{C}$ is simply the temporal operator *chop*.

This approach seems attractive for handling conventional programming languages since one is not restricted to reasoning about complete programs. However, in the case of Tempura the distinction between programs and formulas is minimal thus permitting a relation such as $\{S\}w$ to be readily expressed as the implication $S \supset w$.

10.3.6 Synthesis from temporal logic

Manna and Wolper [30] investigate techniques for automatically synthesizing CSP synchronization code from temporal logic specifications. This holds much promise since the design of correct routines for interprocess synchronization is generally regarded as tedious and error-prone. One example considered is Dijkstra's well known dining philosophers' problem. Another system consists of a synchronizer S that regulates the activity of two other processes P_1 and P_2 by ensuring that they never simultaneously operate in their respective critical regions. Let us look at how this is handled.

The behavior of each process P_i is expressed in temporal logic as a conjunction of the form shown below:

$$S!begin_i \wedge \Box(S!begin_i \supset \bigcirc S!end_i) \wedge \Box(S!end_i \supset \bigcirc S!begin_i).$$

Here the construct $S!begin_i$ represents a request by process P_i to enter its critical region. Similarly, the construct $S!end_i$ is used when P_i is ready to leave the critical region. Thus, the specification states that each process initially makes a request to enter its critical region. Furthermore, whenever it enters the region, it subsequently exits and whenever it exits it subsequently makes a request to reenter.

The specification of S is a conjunction of two formulas. The first requires that whenever P_1 enters its critical region, P_2 is not allowed in its own critical region until after P_1 exits. This is expressed in temporal logic in the following way:

$$\Box \left[P_1?begin_1 \supset ([\neg P_2?begin_2]\,\mathcal{U}\,[P_1?end_1]) \right].$$

The second part of the conjunction is similar but reverse the roles of P_1 and P_2. The *until* operator $\mathcal{U}$ provides a means of resolving conflicting requests.

Note that the usage of the operators ! and ? differs slightly from that of the version of CSP described previously since process names are given instead of channels. This convention is in fact adapted from an earlier variant of CSP described by Hoare in [21].

Given the specifications just described, Manna and Wolper show how to mechanically produce a set of finite-state automata that satisfy the various constraints. The authors then derive CSP programs in a straightforward manner. The synchronizer S has the following process associated with it:

$$*[N = 1;\ P_1?begin_1 \rightarrow N := 2$$
$$[]\ N = 1;\ P_2?begin_2 \rightarrow N := 3$$
$$[]\ N = 2;\ P_1?end_1 \rightarrow N := 1$$
$$[]\ N = 3;\ P_2?end_2 \rightarrow N := 1].$$

The operator ∗ indicates unlimited sequential repetition of the associated statement. The body of the system consists of four transitions separated by the operator []. This specifies that they are to be nondeterministically selected whenever their respective guard conditions are enabled. The variable N is used to maintain the state of the synchronizer. For example, the transition given below is enabled when N equals 1 and P_1 wishes to enter its critical region:

$$N = 1;\ P_1?begin_1 \to N := 2.$$

The process P_1 has the following CSP implementation:

$$^*[N = 1;\ S?begin_1 \to N := 2$$
$$[\!]\ N = 2;\ S?end_1 \to N := 1].$$

The program for process P_2 is similar:

$$^*[N = 1;\ S?begin_2 \to N := 2$$
$$[\!]\ N = 2;\ S?end_2 \to N := 1].$$

Even though CSP is the target language used in this work, we imagine that similar techniques could be applied to synthesizing Tempura programs. One resulting advantage would be the ability to go from high-level specifications to implementations without leaving temporal logic.

10.3.7 Automatic verification of circuits

Mishra and Clarke [33] use a temporal logic called CTL in a system that automatically verifies asynchronous digital circuits. The system accepts a behavioral specification given in CTL and generates a state-transition graph from it. This graph acts as a model against which various temporal properties can be checked. The graph can also be viewed as an implementation of the original specification. The main example presented is a self-timed queue element containing signals for passing data as well as for performing handshaking. Perhaps this kind of verifier can be modified to handle suitable hardware specifications given in ITL.

10.4   Conclusions

The present work has investigated Tempura, a programming language based on Interval Temporal Logic. The ITL formalism provides a way to treat such programming concepts as assignment and loops as formulas about intervals of time. Therefore, Tempura programs, their specifications and their properties can all be expressed in the same formalism. Furthermore, this approach provides a unified way for modelling both hardware and software. In the future, we hope to gain more experience with using ITL and Tempura to simulate and reason about descriptions of hardware devices and other types of parallel systems. In addition, we plan to explore the feasibility of using Tempura as a general-purpose programming language.

Bibliography

[1] E. A. Ashcroft and W. W. Wadge. Lucid: A formal system for writing and proving programs. *SIAM Journal of Computing 5*, 3 (September 1976), 336-354.

[2] E. A. Ashcroft and W. W. Wadge. Lucid, a nonprocedural language with iteration. *Communications of the ACM 20*, 7 (July 1977), 519-526.

[3] H. Barringer, R. Kuiper and A. Pnueli. Now you may compose temporal logic specifications. In *Proceedings of the Sixteenth Annual ACM Symposium on Theory of Computing*, Washington D. C., April, 1984.

[4] G. Berry and L. Cosserat. The Esterel synchronous programming language and its mathematical semantics. Technical report, Ecole Nationale Supérieure des Mines de Paris (ENSMP), Centre de Mathématiques Appliquées, Valbonne, France, 1984.

[5] R. Burstall. Program proving as hand simulation with a little induction. In *Proceedings of IFIP Congress 74*, North Holland Publishing Co., Amsterdam, 1974, pages 308-312.

[6] A. Chandra, J. Halpern, A. Meyer, and R. Parikh. Equations between regular terms and an application to process logic. *Proceedings of the Thirteenth Annual ACM Symposium on Theory of Computing*, Milwaukee, Wisconsin, May, 1981, pages 384-390.

[7] K. L. Clark and S. Gregory. Parlog: Parallel programming in logic. Technical report DOC 84/4, Department of Computing, Imperial College, London, April, 1984. To appear in *ACM Transactions on Programming Languages and Systems*.

[8] K. L. Clark, F. G. McCabe and S. Gregory. IC-Prolog language features. In K. L. Clark and S.-Å. Tärnlund, editors, *Logic Programming*, pages 253–266, Academic Press, London, 1982.

[9] W. F. Clocksin and C. S. Mellish. *Programming in Prolog*. Springer-Verlag, Berlin, 1981.

[10] A. Colmeraurer, H. Kanoui, R. Pasero and P. Roussel. Un système de communication homme-machine en Français. Research report, Groupe de Recherche en Intelligence Artificielle, Université d'Aix-Marseille, France, 1973.

[11] H. B. Enderton. *A Mathematical Introduction to Logic*. Academic Press, New York, 1972.

[12] M. J. C. Gordon. LCF-LSM: A system for specifying and verifying hardware. Computer Laboratory, University of Cambridge, Technical report 41, 1983.

[13] R. W. S. Hale. Modelling a ring network in interval temporal logic. To appear in *Proceedings of Euromicro 85*, Brussels, Belgium, September, 1985.

[14] J. Halpern, Z. Manna and B. Moszkowski. A hardware semantics based on temporal intervals. In *Proceedings of the 10-th International Colloquium on Automata, Languages and Programming*, Number 154 in the series *Lecture Notes in Computer Science*, Springer-Verlag, Berlin, 1983, pages 278–291.

[15] D. Harel. *First-Order Dynamic Logic*. Number 68 in the series *Lecture Notes in Computer Science*, Springer-Verlag, Berlin, 1979.

[16] D. Harel, D. Kozen, and R. Parikh. Process logic: Expressiveness, decidability, completeness. *Journal of Computer and System Sciences 25*, 2 (October 1982), pages 144–170.

[17] E. C. R. Hehner. Predicative programming (parts I and II). *Communications of the ACM 27*, 2 (February 1984), pages 134–151.

[18] P. Henderson. *Functional Programming: Application and Implementation*. Prentice-Hall International, London, 1980.

[19] C. A. R. Hoare. An axiomatic basis for computer programming. *Communications of the ACM 12*, 10 (October 1969), pages 576–580, 583.

[20] C. A. R. Hoare. Towards a theory of parallel programming. In C. A. R. Hoare and R. H. Perrott, editors, *Operating Systems Techniques*, pages 61–71. Academic Press, London, 1972.

[21] C. A. R. Hoare. Communicating sequential processes. *Communications of the ACM 21*, 8 (August 1978), 666-677.

[22] C. A. R. Hoare. *Communicating Sequential Processes*. Prentice Hall International, London, 1985.

[23] C. J. Hogger. *Introduction to Logic Programming*. Academic Press, Orlando, Florida, 1984.

[24] Inmos Ltd. $Occam^{TM}$ *Programming Manual*. Prentice Hall International, London, 1984.

[25] R. Kowalski. Predicate logic as programming language. In *Proceedings of IFIP Congress 74*, North Holland Publishing Co., Amsterdam, 1974, pages 569–574.

[26] R. Kowalski. *Logic for Problem Solving*, Elsevier North Holland, Inc., New York, 1979.

[27] J. McCarthy, P. W. Abrahams, D. J. Edwards et al. *Lisp 1.5 Programmer's Manual*. MIT Press, Cambridge, Massachusetts, 1962.

[28] Z. Manna and A. Pnueli. Verification of concurrent programs: The temporal framework. In R. S. Boyer and J. S. Moore, editors, *The Correctness Problem in Computer Science*, pages 215–273, Academic Press, New York, 1981.

[29] Z. Manna and A. Pnueli. How to cook your favorite programming language in temporal logic. In *Proceedings of the Tenth Annual ACM Symposium on Principles of Programming Languages*, Austin, Texas, January, 1983, pages 141–154.

[30] Z. Manna and P. L. Wolper. Synthesis of computing processes from temporal logic specifications. *ACM Transactions on Programming Languages and Systems 6*, 1 (January 1984), pages 68–93.

[31] G. J. Milne. CIRCAL and the representation of communication, concurrency and time. *ACM Transactions on Programming Languages and Systems 7*, 2 (April 1985), pages 270–298.

[32] R. Milner. *A Calculus of Communicating Systems*. Number 92 in the series *Lecture Notes in Computer Science*, Springer-Verlag, Berlin, 1980.

[33] B. Mishra and E. M. Clarke, Automatic and hierarchical verification of asynchronous circuits using temporal logic. Technical report CMU-CS-83-155, Department of Computer Science, Carnegie-Mellon University, September, 1983.

[34] B. Moszkowski. *Reasoning about Digital Circuits*. PhD Thesis, Department of Computer Science, Stanford University, 1983. (Available as technical report STAN–CS–83–970.)

[35] B. Moszkowski. A temporal analysis of some concurrent systems. To appear in *Analysis of Concurrent Systems*, in the series *Lecture Notes in Computer Science*, Springer-Verlag, Berlin.

[36] B. Moszkowski. A temporal logic for multilevel reasoning about hardware. *Computer 18*, 2 (February 1985), pages 10–19.

[37] B. Moszkowski and Z. Manna. Reasoning in interval temporal logic. In *Proceedings of the ACM/NSF/ONR Workshop on Logics of Programs*, Number 164 in the series *Lecture Notes in Computer Science*, Springer-Verlag, Berlin, 1984, pages 371–383.

[38] A. Pnueli. The temporal semantics of concurrent programs. *Theoretical Computer Science 13*, (1981), pages 45–60.

[39] V. R. Pratt, Semantical considerations on Floyd-Hoare logic. In *Proceedings of the Seventeenth Annual IEEE Symposium on Foundations of Computer Science*, Houston, Texas, October, 1976, pages 109–121.

[40] N. Rescher and A. Urquhart. *Temporal Logic*. Springer-Verlag, New York, 1971.

[41] R. L. Schwartz, P. M. Melliar-Smith, and F. H. Vogt. An interval logic for higher-level temporal reasoning. In *Proceedings of the Second Annual ACM Symposium on Principles of Distributed Computing*, Montreal, Quebec, Canada, August, 1983, pages 173–186.

[42] E. Y. Shapiro. Systems programming in Concurrent Prolog, In *Proceedings of the Eleventh Annual ACM Symposium on Principles of Programming Languages*, Salt Lake City, Utah, January, 1984, pages 93–105.

[43] D. E. Shasha, A. Pnueli and W. Ewald. Temporal verification of carrier-sense local area network protocols. In *Proceedings of the Eleventh Annual ACM Symposium on Principles of Programming Languages*, Salt Lake City, Utah, January, 1984, pages 54–65.

[44] C. Tang. Toward a unified logic basis for programming languages. In *Proceedings of IFIP Congress 83*, Elsevier Science Publishers B.V. (North-Holland), Amsterdam, 1983, pages 425–429.

[45] W. W. Wadge and E. A. Ashcroft. *Lucid, the Dataflow Programming Language*. Academic Press, Orlando, Florida, 1985.